# The Ecology of Neotropical Savannas

# The Ecology of Neotropical Savannas

**Guillermo Sarmiento**

Translated by Otto Solbrig

Harvard University Press

Cambridge, Massachusetts, and London, England   1984

*Library of Congress Cataloging in Publication Data*

Sarmiento, Guillermo.
    The ecology of neotropical savannas.

    Translation of: Estructura y funcionamiento de
    sabanas neotropicales.
    Bibliography: p.
    Includes index.
    1.   Savanna ecology—Venezuela.
    2.   Savanna ecology.
I.  Title
QH130.S2713   1984     574.5'2643'098     83-12904
ISBN 0-674-22460-4

To the tragical and generous *campesinos* of Venezuela,
the true ecologists of this country,
whom with humility and respect we view as our teachers

# Author's Acknowledgments

The studies that lay the foundation for this book were financed in part by grants of the Consejo Nacional de Investigaciones Científicas y Tecnológicas (CONICIT) of Venezuela, and by a grant of the Consejo de Desarrollo Científico y Humanístico de la Universidad de los Andes. To both institutions I am very grateful. Much of the work was financed by the Grupo de Ecología Vegetal of the Department of Biology of the Faculty of Sciences of the Universidad de los Andes, and I am glad to acknowledge their valuable support.

I especially wish to thank the technical personnel of the Laboratory of Plant Ecology. Hector Molina shared many field trips and Luis Nieto was always a most valuable collaborator. Gladys Lobo typed the manuscript with dedication and good will. To all, my warmest thanks.

I consider this book a still-provisional summary of many years of research in the Venezuelan savannas. My contact with this ecosystem began in 1966 and since then there have been few occasions when I have not planned, discussed, and performed the research with Dr. Maximina Monasterio. Consequently, I consider this work to be a successful long-term cooperative effort. Without this collaboration the conclusions would have been more meager and the task not so pleasant.

G.S.

# Translator's Preface

Tropical savannas are home to a large part of the population of the world. They constitute one of the most important ecosystems from an economic, social, and scientific point of view. Savannas are particularly extensive in South America and Africa. Despite their great importance, savanna ecosystems have received less attention than tropical rain forests. Especially needed are integrated studies on the function of these systems. Studies on tropical savannas have provided many descriptions of individual species, but synthetic approaches that stress the main functional pathways are still missing.

Guillermo Sarmiento has been studying neotropical savannas for close to twenty-five years, in Argentina, Brazil, and especially Venezuela. In this book Sarmiento develops an integrated view of the functioning of tropical savannas. Making broad use of the literature and his own extensive knowledge of tropical savannas, Sarmiento discusses the importance of phenology, water economy, production, and the always controversial role of nutrient economy in the functioning of the savanna.

Professor Sarmiento kindly read and corrected the draft of the translation. Lisa Betteridge prepared the line drawings and Elizabeth Maynard provided bibliographical assistance. To all who have helped me, I am greatly indebted.

O.T.S.

Arlington, Massachusetts
October 1983

# Contents

# The Ecology of Neotropical Savannas

# 1

# The problem of the tropical savanna

The savanna ecosystems play a major role in the configuration of the natural landscapes and the economic life of vast areas in the tropical regions of the New and Old worlds. In some countries these grass-dominated ecosystems are the principal biotic resource. The efficient management of these grasslands will determine to a great extent whether the standard of living of the rural population can be improved.

Interest in savannas has increased in recent years, both interest in their basic ecological aspects and interest in developing management systems that may allow a better use of their agricultural potential. A "State of Knowledge" report prepared by UNESCO in 1979 discussed the characteristics of tropical grazing lands from a very broad perspective, including consideration of the patterns of human occupation and the different types of human use. Despite the great value of this survey, in addition to savannas it covers all the other tropical grazing-land ecosystems. Consequently it is difficult to identify the characteristics that are representative of and unique to savannas as here defined. Moreover, the report makes clear that much less is known about the American tropics than about other tropical regions. The few available works on the neotropical savannas are constantly cited. One useful work (Huntley and Walker 1982) concentrates on Africa, particularly on Nylsvley, a South African savanna, where notable progress in understanding the system has been achieved. Unfortunately, in spite of the evident convergences between the South African and the American savannas, they represent very different ecological systems. The work of Bourlière (1983) probably represents the most important and complete survey published to date on tropical savannas, especially concerning consumers and secondary production, but here also there are constant references to the lack of information on American savannas.

Savannas constitute approximately a third of the surface of Venezuela. They are found in all the natural regions of the country, but dominate the landscape in the plains that surround the Orinoco River, the so-called *llanos*, forming what from every point of view is a "savanna country." The llanos between Colombia and Venezuela cover a surface of approximately half a million square kilometers, constituting the largest uninterrupted surface of neotropical savanna north of the Equator.

In the center of South America, in the area of the old Precambrian shield, and occupying most of the *planalto* of interior Brazil, savannas occupy millions of square kilometers. Phytogeographers call this formation either province of the cerrado (Cabrera and Willink, 1973) or formation of the cerrado (Eiten, 1977). If we include neighboring territories where savannas dominate the landscape, such as the llanos of Moxos in eastern Bolivia, which extend to the very foot of the Andes, or the Gran Pantanal, which extends southward to the headwaters of the Paraguay River basin, the central savanna region of South America acquires subcontinental proportions (Sarmiento, 1983).

Even the Amazonian region, the gigantic "land of rain forests," has islands of herbaceous vegetation that like an archipelago connect the lowland areas of the cerrado with the llanos. The Precambrian area of the Guayanas shield also has extensive regions of savannas with isolated fingerlike extensions to the Atlantic coast forming an arc between Guyana and the mouth of the Amazon (fig. 1). Central America and the islands of the Antilles contain savanna areas of varying dimensions; in Cuba, the largest of the Caribbean islands, the savanna constitutes a landscape so distinctive that it gave rise to the name "savanna" (Sarmiento, 1982).

The origin, age, nature, and dynamics of neotropical savannas, as well as the environmental factors that determine or condition savanna ecosystems, have been and continue to be debated. Many contradictory interpretations have been advanced, including the extreme view that all these systems had a recent origin due to human activity. As the ecological knowledge of tropical savannas and especially of neotropical savannas increases, the accumulation of new data produces more adequate interpretations. But the very

**Figure 1** Principal savanna regions in Central and South America: (1) cerrado; (2) llanos of Moxos; (3) llanos of Orinoco; (4) gran sabana; (5) savannas of the Río Branco-Rupununi; (6) coastal savannas of the Guayanas; (7) Amazonian campos; (8) llanos of the Magdalena; (9) savannas of Miskito. (Modified from Sarmiento and Monasterio, 1975.)

proliferation of data obtained in geographically separated and ecologically distinct areas often leads to considerable confusion, which makes it desirable and urgent to synthesize the available information. Even if the resulting works produce only provisional theories and conclusions, they will have heuristic value in the search for future research directions and approaches that will lead to increasingly firmer conclusions. The primary objective of this book is therefore to increase understanding of the ecosystems of neotropical savannas through the presentation and discussion of their structural and functional aspects. In this field of research scientists frequently work in teams and depend a great deal on the work and data of colleagues whose work complements their own. I have also considered my work to be part of the collective endeavor, and include even the reader in some measure as participant in this effort; hence my preference for the pronoun "we" in all the discussions that follow.

Not only are the Venezuelan savannas the ones that the author knows best, but they appear also to be the best analyzed in South America from an ecological standpoint, although researchers in other South American countries have made considerable contributions, in particular the pioneering work of the school founded by F. Rawitscher in São Paulo, Brazil (see references in Ferri, 1963, 1977, and Labouriau, 1966). Aside from some general works, including biogeographical studies and those concerned with the broader correspondences between climate, soils, relief patterns, vegetation, and land use (see bibliography in Huber, 1974, and Medina and Sarmiento, 1979), the most substantial investigations done on Venezuelan ecosystems have been carried out until now in a few places in the llanos, indicated in Figure 2. It is in these localities that most quantitative studies of composition, structure, phenology, production, water balance, and nutrient economy have been undertaken. The Venezuelan savannas will therefore be the center of interest in this book, but we will always keep in mind the wider continental picture, and whenever possible make comparisons with the better-known African savannas. Likewise, we will try to establish comparisons with the temperate grasslands, which have many structural and functional characteristics in common with tropical savannas, and comparisons with the tropical rain forest, the other half of the ecological world of the humid warm tropics.

At the outset it is important to define as unequivocally as possible what a tropical savanna is, in order to eliminate any imprecision in usage of the term. Then we will explain briefly why a "tropical savanna problem" exists within ecology. In the following chapters we will examine the structural and functional phenomena that characterize savanna ecosystems. Chapter 2 focuses on aspects of spatial organization or architecture, and Chapter 3 analyzes the functional rhythms of the ecosystems and of each of their component parts throughout the year, since it is these systems, within the

**Figure 2** Principal research sites in the Venezuelan llanos. The *dot-dashed* line indicates the approximate limit of the savanna region.

tropics, that show the most surprising changes. Chapter 4 will discuss all those phenomena that are related to primary production, and in the following two chapters, first the water balance and then the economy and cycle of the principal nutrients. The final chapter consists of discussion, within a very broad context, of the conclusions arrived at in the previous chapters. Let us turn now to the definition of savanna in the sense that it will be used in this book.

## What is a tropical savanna?

The word *sabana* is of Amerindian origin. In pre-Columbian times it was used in Haiti and Cuba to designate plains devoid of trees but with a tall herbaceous cover (Lanjouw, 1936, cited by Beard, 1953). Today it is used in ordinary Spanish in almost all countries of Spanish America to refer to a flat, grassy landscape that may or may not have some isolated shrubs or trees. It is used in Venezuela in opposition to *mata* (an isolated forest grove or forest island within an open landscape), or *montaña* (forest in popular usage), — that is to say, in contrast to forested landscapes where tall grasses do not play an important role. Note that the word has two meanings: in one sense it refers to a flat and open landscape and in another it refers to the vegetation that characterizes that landscape. Furthermore, there is usually the connotation of grazed land, because in the Americas the savannas have been used principally for grazing.

The Portuguese language did not incorporate the same word, utilizing instead the word *campo* to designate savanna landscape and vegetation. In Brazil campo is applied to the tropical open grassland formations (the *campos cerrados* in the broad sense) and also to temperate and subtropical grasslands more akin to the *pampas* of the La Plata river basin. The word campo is used in Spanish-speaking countries of southern South America in the same sense as in Brazil, though in the subtropical regions of Argentina both campo and savanna are used.

In the scientific literature the word savanna has been acquiring an ever broader meaning, to the point that it is used, both in the Americas and in other parts of the world, for any formation that is not a forest or a semidesert. The word in this sense has come into German, English, and French, and its use has been extended throughout the world. So, for example, for Dansereau (1957) savanna is a mixed physiognomy of grasses and woody plants and as such constitutes, along with the forest, the grassland, and the desert, one of the four groups, or biochores, into which he classifies all possible types of vegetation. Thus the original meaning of savanna has been largely lost, and what remains is a sense of savannas as consisting of a mixture of grasses and woody plants. Some German geographers such as Troll (1950) and Lauer (1952) extended the term even further and applied it to any type of vegetation between a forest and a desert, thereby following Jaeger (1945), who proposed savanna as the name for a climatic zone within the tropical and subtropical zones — that is, the climatic zone of summer rains, in the astronomical sense of summer, the only one that can have any meaning in the tropics.

Given the broad and imprecise use of this word in the scientific literature, some authors have proposed that its use should be avoided altogether. Nevertheless, with the qualification of "tropical," the savanna has been recognized as one of the most characteristic tropical landscapes in all continents (Schimper and Von Faber, 1935; Walter, 1971). To summarize, savanna has been used to describe the following formations:

1. Open formation (not forested) dominated by grasses, in the low latitude tropics, where trees and shrubs, if present, are of little physiognomic significance (original meaning of the word and popular meaning of the word in Spanish-speaking countries). (Following Beard, 1953, we will use the word in this sense but will elaborate a more precise definition.)
2. Mixed tropical formation of grasses and woody plants, excluding pure grasslands (Walter).
3. Mixed formation in any geographical area; a purely physiognomic concept (Dansereau).
4. Any formation or landscape within the region with summer rains, that is, within the hydric regime called "tropical" (Jaeger, Troll, Lauer).

The word savanna can only be usefully employed in ecological discussions if its meaning is clearly stated and if it is not used in a purely physiognomic sense, because from a purely descriptive point of view it is better to use words with a broader meaning such as grassland, woodland, scrubland, and so on. Since the beginning of our work we have reserved the word savanna for a type of tropical vegetation where certain forms of grasses dominate and where seasonal droughts and frequent fires are normal ecological factors. That is, we called savannas those formations that are similar to the ones to which the word was originally applied, but included in our definition of savannas more important ecological characteristics, as well as their physical appearance.

We now wish to complete and extend the definition of savanna that we used in previous publications (see, for example, Monasterio and Sarmiento, 1968; Sarmiento and Monasterio, 1971), which was based, as already mentioned, on the explicit definition of Beard (1953). Our previous definition was restricted to the physiognomic and ecological characteristics of the vegetation, and did not explicitly include the habitat factors characteristic of the tropical climate. We believe that one must move from viewing the tropical savanna as a unit of vegetation to considering the tropical savanna as an ecosystem. As later chapters will show, there are many different types of savanna ecosystems. Nevertheless, it is possible to define an ecosystem-type for the tropical savanna that encompasses all variants. Taking as a base the definition we used before and broadening it, we will characterize a tropical savanna as a type of ecosystem of the warm (lowland) tropics dominated by a herbaceous cover consisting mostly of bunch grasses and sedges that are more than 30 cm in height at the time of maximum activity. The herbaceous cover shows a clear seasonality in its development, with a period of low activity related to water stress (xeropause, in the sense of Stanyukovich, 1970). Fire in these systems is a recurring natural factor, and fires may also be started by people once a year.

The savanna may include woody species (shrubs, trees, palm trees), but they never form a continuous cover that parallels the grassy one.

This very broad definition applies to a diversity of ecosystems that differ among themselves in many important characteristics. Consequently, we will be establishing refinements and subdivisions as we go along. Nevertheless, all ecosystems included within this definition share a number of basic characteristics that allows their treatment as a unit, at least in a first analysis.

This definition excludes a number of ecosystems that one or another author considers to be savanna, for example, forested formations where trees form a continuous canopy and where the herbaceous cover is meager or nonexistent; open grasslands such as steppes; the grasslands formed by low grasses such as meadows, lawns, and so on; herbaceous formations dominated by families other than the Gramineae or Cyperaceae (wetland ecosystems); grassland formations without seasonality (such as boglands); and any ecosystems outside the lowland tropics. In this regard it is important to note that the qualification *tropical* applied to these ecosystems does not have only a geographical meaning, in which case it would be of little interpretative value; it also implies that a number of processes and conditions are operating at the ecosystem level that are restricted to the tropical ecological zone. That is, limiting these systems to the warm tropics emphasizes the ecological peculiarities of this environment and their influence on all the other conditions of such ecosystems, thereby specifying that we are dealing with a set of interrelations between biotic and abiotic components that, although common to all ecosystems, acquire in the savanna systems unique characteristics. The most essential functional processes of these ecosystems are uniquely tropical: their productive processes, their seasonal cycles and rhythms, the environmental stresses to which their species are subjected, their mineral cycles, and their water balances. In this sense tropical savannas might differ more from physiognomically similar ecosystems of the temperate zone than from other tropical ecosystems. With this definition of savanna established, we go on to consider why further research on savannas is needed.

# Need to explain the existence of savannas

From the beginning of the nineteenth century, biogeography has developed following the major biological explorations of those parts of the world which were previously unknown to the Western world, among them the tropical regions. The results of these explorations produced the first global phytogeographical sketches, outlining the typology of the vegetation and describing its distribution. Beginning in the twentieth century, studies of climates and of soils were undertaken from a world perspective, and their findings helped the rapid development of ecology as the field where biogeography, climatology, and soil science intersect. From this intersection was born the concept of world vegetation types and of biomes, what today is better and more appropriately called ecosystem types.

To explain the geographical distribution of biomes, ecologists repeatedly referred to the concept of climatic zonation, and on this concept rests the conceptual framework of succession and biotic climax theory (Clements, 1916; Weaver and Clements, 1929). According to this conceptual scheme, which was widely accepted for many decades, each climatic region has a unique stable, or "climax," biotic formation, the climatic climax. If the geographical distribution of a particular plant formation does not correspond with any given climatic zone, that formation is considered to be "azonal" or "intrazonal"; examples of azonal formations are seral or successional formations, in temporary disequilibrium with the climatic conditions of the environment, and an example of an "intrazonal" formation is a disclimax formation, whose transformation toward a climax is temporarily stopped by environmental factors (particularly edaphic ones) linked to a very distinctive kind of environment.

According to this ecological interpretation of the distribution of vegetation, woody formations were considered without question to be the climatic climaxes under all conditions, except in those cases where a major environmental factor such as temperature or humidity, made it impossible for woody formations to become established. Except in arid or semiarid climates, or arctic or alpine environments, the supremacy of woody formations was not questioned.

This generalized concept of the dominance of forests arose for two reasons. First, there is a psychological factor: in the temperate zones (from the limits of the subtropical regions to the artic), where the majority of biogeographers get their training, the forest, in any of its forms and variants, is undoubtedly the primitive climax formation, to which the natural evolution of the vegetation tends when not subjected to human influences. Second, it must be mentioned that, with few exceptions, when a woody formation has become established in a site, it will dominate the shrubby and herbaceous forms, because of its control over climatic factors.

As a natural corollary of these more or less explicit basic concepts of the biogeographical literature, all herbaceous or nonforested formations of the temperate or tropical regions whose existence could not be explained on climatic arguments were logically regarded as azonal. Thus the problem arose of trying to explain the existence of natural, nonforested formations in climates that were neither too dry nor too cold to sustain a continued canopy forest. Many widely diverse explanations were evolved to justify the existence of the North American prairies and the South American pampas, which could not be considered to be semiarid steppes and which to the contrary occupied some of the most promising agricultural land of the world, as well as the existence of the savannas and grasslands of other nonarid tropical and subtropical regions. When edaphic constraints were not in evidence, or the presumably seral nature of a formation was difficult to prove, paleogeographical explanations were adduced or certain irreversible environmental modifications tied to the activity of early man were brought forward. Eventually, the lack of trees was attributed to fire, an ecological factor that had been underestimated until then, and that periodically consumed the herbaceous vegetation of many of these places.

Within this broad problem of the etiology of all grassland formations, those questions specific to the tropical savannas can be formulated thus: How can the apparently indefinite existence of a savanna in a given region of the tropics be explained when it is established that trees can grow in areas either wetter or drier than the savannas, and that forests are well established in climates that are less seasonal than those of savannas, as well as in climates that are more seasonal?

To solve this problem several alternative explanations (Sarmiento and Monasterio, 1975), including the following, have been proposed.

1. Tropical savannas are maintained only under special climatic conditions; that is, there exists a special savanna climate that favors this type of formation over any type of forest. However, since it is difficult to characterize this savanna climate with the usual elements that are used to define a given climatic type (mean temperature, rainfall, and so on), more subtle variables must be used, ones that will demonstrate the factors that separate this climate from forest climates. Once the savanna climate has been characterized, the theoretical correspondence between plant formation and climatic-type climax can be established, and there is no need to use additional explanations that might contradict the generally accepted theory.

2. Tropical savannas are due to edaphic factors that do not allow the maintenance of a forest. Viewed this way savannas would be intrazonal formations; the climax theory would therefore still be valid. The problem becomes one of identifying the edaphic factors that favor savannas.

3. Recurrent fires do not allow the rebuilding of an ancient forest that has been destroyed. This hypothesis presents additional problems, such as explaining what that primitive forest may have been, how it could have been destroyed over such vast areas, and how fire can impede its reestablishment.

4. Environmental conditions in savanna areas are marginal for the establishment of forests. An unstable equilibrium is therefore postulated, one that can easily move toward the establishment of herbaceous vegetation as the consequence of human disturbance. The theory of the climax forest is sustained, even though the explanation is somewhat contrived.

5. The biogeochemical cycle of the wet tropics leads after a certain time to conditions that favor savannas, which are stable until a new geomorphogenetic and climatic cycle is initiated. The factors responsible for the existence of savannas are pedological and chemical in nature, but their origin falls within the general

framework of morphogenesis in tropical climates. Consequently, savannas would be postclimax formations covering old landscapes.

These hypotheses illustrate some of the explanations advanced by different authors to justify the presence of nonforest formations in the wet tropics. They confirm the existence of a problem in accounting for the existence of tropical savannas that has not yet been resolved to everyone's satisfaction. Now we will leave this crucial subject until the last chapter of the book, when after having analyzed the principal ecological processes of the tropical savannas, we will be able to discuss more adequately these important theoretical issues in tropical biology.

# 2

# The architecture of
# the savanna

This chapter will consider the spatial distribution of the biomass in the savanna ecosystems, both its aerial, or epigeal, parts and the underground, or hypogeal, ones. But examining the spatial dimensions of the biomass gives only a static vision of the ecosystem, a single cut through time; so it is essential to specify what part of the yearly cycle a given spatial organization corresponds to. This is true for any type of ecosystem, but in the savannas the cyclical changes are more pronounced than in other areas, and a substantial part of the biomass is renewed every year. Here more than in other regions the structure of the biomass at any one time depends on preceding events, as for example, whether and when it was burned, whether it has been grazed and with what intensity, and so on.

The precise characterization of the spatial structure of the biomass will allow us to establish the architecture of the ecosystem. It will give us the basis for understanding the seasonal changes that take place in every developmental stage and for quantifying the effective occupation of the habitat by the vegetation, which constitutes the principal building element of an ecosystem. Additionally, by establishing the spatial distribution of the various functional compartments, such as the assimilatory biomass, reproductive biomass, standing dead biomass, and others, we will be able to find and describe the sites where the basic ecological processes are taking place, contributing in this way to a better understanding of the physiological processes and of the micro-environmental factors that regulate growth. Generally speaking, the structural aspects that correspond to the vertical plane and those that reflect the spatial heterogeneity in the horizontal plane are considered separately. The vertical structure of a community has commonly been described in terms of various strata where the concentration of the biomass reaches relative maxima. Each stratum can be characterized not only by its amount of cover and its height, or depth if the level is hypogeous, but also by the relative percentage of the total biomass that it represents, as well as by its morphoecological characteristics and specific composition.

This stratification of the vegetation arises from competition between its species for the resources of the environment, and from interaction with environmental gradients, especially temperature and humidity in the lower level of air that is in direct contact with the ground. Below ground the principal environmental gradients are determined by the vertical distribution of water and nutrients and the concentrations of oxygen and carbon dioxide in the soil atmosphere. Once the vegetation forms some stable structure, it creates new gradients that reinforce the existing habitat stratification. In this manner a complex bilateral correspondence is established between the spatial variations of the plant biomass and environmental factors.

In the neotropical savannas one can initially differentiate strata formed fundamentally by herbaceous plants from those integrated by woody elements, whether shrubs or trees. Each of these categories may be subdivided on the basis of height, cover, and so on. The epigeous plant biomass in the savannas has conventionally been divided into four different strata: two formed by woody plants, and two by grasses (Beard, 1953; Monasterio and Sarmiento, 1968; Sarmiento and Monasterio, 1971). But our definition of tropical savanna eliminates three of these strata; the remaining stratum, the upper herbaceous, is the only essential one, since its presence is sufficient to characterize a savanna ecosystem.

The vertical differentiation of strata, although sometimes useful in a first morphoecological and structural description of some ecosystems, does not fit the reality of a savanna ecosystem when the vertical structure is established on a more quantitative basis. In the savanna more or less continuous distribution of the biomass over space obtains, with gradual changes in height, within characteristic gradients for each plant formation. We will thus use the strata nomenclature to formulate an initial qualitative description, and then move, on the basis of the meager data available, to a more flexible structural interpretation based on quantitative determinations.

In addition to the fairly discontinuous distribution of the vertical biomass in a community that is called stratification, there is also the clustering of unities, whether individuals or life forms, in more or less specific places in the terrain, constituting the elements of a horizontal structure. That is, within the two coordinates of a horizontal plane a community may present a certain anisotropy, so that it can be considered in its totality as a kind of mosaic that is interpenetrated by two or more structurally different elements. For example, some savannas have a herbaceous cover interspersed by groups of woody plants, which in some cases may be isolated individuals of a single species. The relatively continuous herbaceous matrix constitutes one strutural element, while the groups of woody plants constitute a different, discontinuous structural unit that may cause changes in the contiguous herbaceous element, in addition to creating ecological niches and distinctive microhabitats that can be utilized by different elements of the fauna.

The distinction between structural elements of a given plant community and complex patterns where two or more structurally different vegetation units intermix in the form of a mosaic is not clearcut; it is useful for certain types of analysis but cannot be sustained in many intermediate situations between the extremes.

# Vertical structure

## The herbaceous stratum

In the great majority of the American savannas the herbaceous plants, at their highest level of development, cover the greatest extension of ground, forming the ecologically dominant stratum of the ecosystem. The major exceptions to this generalization are the closed and forested savannas that occur in the region of the cerrado of the Brazilian planalto. Despite a relatively continuous cover of the herbaceous elements and a much more discontinuous one of trees, in many forested savannas (closed tree savannas) trees can form the major percentage of the above-ground biomass. However, even in these cases it is the herbaceous stratum that regulates and conditions the fundamental ecological processes in the ecosystem, such as regrowth after fires, the water balance, annual productivity, mineral cycling, and so on.

In most places in the tropical American savannas the herbaceous stratum reaches in its vegetative state a height of 50 to 100 cm, and at the time of the development of inflorescences up to 120 to 150 cm. Only in some communities that are subjected to river floods does the herbaceous layer grow much higher, as for example in savannas of *Paspalum fasciculatum*. At the other extreme, there are natural grasslands where the grasses form a dense layer not more than 10 to 15 cm in height; these herbaceous formations can hardly be considered savannas, because they lack bunch grasses and have hydromorphic sandy soils, permanently wet, very organic, on the border of lakes and ponds.

The most extensive savannas in tropical America outside of the Brazilian planalto — in particular the great majority of the savannas of northern South America and of Central America — are formed by a herbaceous stratum called "tall herbaceous stratum," although it is only of medium height in comparison with some African savannas where grasses surpass 2 and 3 m. This type of savanna was

**Figure 3** *Trachypogon plumosus-Axonopus canescens* savanna at the Biological Station of the llanos (Calabozo, state of Guarico, Venezuela). The photo was taken two weeks after an accidental fire on 28 December 1968. Among its noticeable effects are the large empty spaces, the reduced basal area of the perennial grasses, the standing remains of the carbonized leaves, the ashes accumulated around the tussocks, as well as the immediate beginning of new growth.

**Figure 4** *Trachypogon vestitus-Paspalum carinatum* savanna on a low sand dune in the state of Apure, Venezuela. The picture was taken at the end of January (middle of the dry season). This is a very dry habitat with a cover of less than 30%. Next to the knife is specimen of *Bulbostylis paradoxa.*

called "tall bunchgrass savanna" by Beard (1953), a name that we may use interchangeably with tall herbaceous stratum.

The total cover of the herbaceous stratum in savannas subjected to fire changes markedly throughout the annual cycle. So, for example, immediately after a fire the burned culms do not cover more than 10 to 20% of the soil surface (fig. 3). Moreover, between the stage of maximum vegetative growth and the end of the annual cycle, plant cover, although ecologically continuous, varies with the type of community and habitat. In the more open communities, such as those that occupy dunes, sandy soils, or soils with surface stones (fig. 4), the herbaceous elements do not cover more than 25% of the soil surface. At the other extreme, in deep soils of medium texture, herbaceous cover often reaches 90 to 100% (fig. 5). To be sure, between these two extremes there are many intermediate situations, depending on the degree of environmental harshness. In any case, savannas that get burned toward the end of the dry period take at least three to four months to regain the point of

**Figure 5** Grass savanna of *Trachypogon plumosus-Axonopus canescens* in a mesa southeast of Calabozo. At this time of greatest aerial development, when the dominant grasses are in full flower and/or fruit, the herbaceous stratum has a cover of 100%.

maximum aerial vegetative development. In savannas that have not been burned for several years, the total cover of the herbaceous stratum varies little during the annual cycle, the only change being in the relative proportions of living and dead biomass.

Grasses, perennial sedges, some annuals, and the great majority of the subfrutescent species are approximately all the same size and determine the maximum median height of this stratum and the level of maximum development of the foliar biomass. However, in all savannas there are smaller annual or perennial species that do not reach beyond 10 or 15 cm in height (fig. 6). These species may be sun-adapted forms that grow and prosper between the tussocks in communities where plant cover is low, or they may be shade-adapted forms that grow in the more humid and cooler microhabitat found below the canopy of tall herbs. The existence of these small species and their relative importance has led to the recognition of a short herbaceous stratum, with an average height of about 10 cm. However, in most savannas the cover and the proportion of the total biomass that this layer represents is so small that it may be ignored in a structural characterization, even though its presence and the species that compose it can provide important ecological and phytogeographic information. Only in certain habitats that have been modified by man do the low sun-adapted species constitute the essential component of the herbaceous communities, but these habitats are not included in our definition of savanna.

Since there were no quantitative data for the spatial organization of the American savannas in the most characteristic stages of their annual cycle, it was necessary to determine the vertical distribution of the epigeal biovolume. The sample chosen was a savanna on the Q4 terrace in the Andean piedmont of the llanos of Barinas (the nomenclature of Tricart and Millies-Lacroix of 1962 identifying the alluvial beds classifies the Andean deposits of Venezuela from Q0 for the recent deposits to Q1 to Q4 for successive Pleistocene geochronological units in order of increasing age). It lay on a

**Figure 6** *Trachypogon vestitus-Leptocoryphium lanatum* in the mesa of Barinas. Picture taken at the end of April, six weeks after the field was burned. The herbaceous stratum is still quite open and there is a fair amount of empty space. *Sporobolus cubensis, Dichronema ciliata,* and *Cissampelos ovalifolia* are perrennial precocious species which are in full bloom, as is *Ruellia geminiflora* (foreground, center), an example of a perennial herb of the low herbaceous stratum that completes its reproductive cycle before the grasses form a closed canopy.

sandy-clay loam, 50 to 60 cm in depth, covering a thick stratum of alluvial pebbles. In this open-tree savanna, where soil texture conditions are quite favorable, the herbaceous layer reaches a maximum cover of 90 to 100%. The stand had been burned before the first rains in April. Sampling by means of the interception method was done at the end of the rainy season (November), when development of reproductive structures was practically complete. One hundred readings of the vegetation were collected with a graduated metal rod placed vertically at random points within a homogeneous 100 m² quadrat. The aerial green, dry, and total biomasses were determined in five randomly selected 1 m² quadrats, while underground biomass was determined from 10 cores that were divided each into 10 cm intervals from the surface to 60 cm below ground, where the rocky underlayer made further sampling impossible.

The data thus obtained made it possible to assess the vertical distribution of the green and dry biovolume and of the vegetative and reproductive structures, and to compare these results with the vertical distribution of the below-ground biomass. The data also provided the means to calculate the leaf surface and the total green area at the time of the measurements and to compare them with the values of the biomass.

The interception method has clear advantages over other methods used previously in tall grasslands; it is more precise and has an acceptable margin of error. Another method, the collection of total harvest and later separation in height classes, introduces errors due to inevitable losses and fragmentation of the material during its harvest, packaging, and later measuring; while the ten-point interception method used in the savannas of Lamto on the Ivory Coast (Poissonet and Cesar, 1972) is considerably more time consuming without much gain in precision, because of the clumping of the data and the complexity of the method.

Calculation of the number of interceptions of the metal rod with all the aerial organs produces a measure of the total biovolume of the vegetation at that time. This measurement cannot be directly translated into biomass, since the biovolume of the green parts corresponds to the fresh weight, while that of the dry parts approximates the oven-dry weight. Consequently there is no direct correspondence between biovolume and biomass, except when all the vegetation is formed either by all green or by only dry tissue.

When measuring the biovolume two sets of records were maintained. One set was recorded by species, and the three dominant species, *Trachypogon vestitus, Axonopus canescens,* and *Leptocoryphium lanatum* were recorded separately from the remaining 26 species, which were all put into one category. A separate set of data was recorded for living standing vegetative organs, dead standing vegetative organs, and reproductive organs, the last category being restricted to a few inflorescences, since the blooming time of the principal species had been over for several months at the time of the study.

Figures 7 and 8 show the total biovolume and the percentages of green, dry, and reproductive biovolume and the proportions corresponding to the three dominant species and the 26 others. The principal conclusions from these data are the following:

1. The total biovolume reaches its maximum value in the lowest stratum (0 – 10 cm) and decreases with height up to 1 m, and then values vary irregularly up to 140 cm. Close to 30% of the total biovolume is found between 0 and 10 cm, and 55% of the total is found between 0 and 20 cm. The decrease of biovolume with height is gradual up to 50 cm and more dramatic after that up to 1 m. The lowest 50 cm contained 92% of the biovolume.

2. No clear stratification of the herbaceous layer was apparent, since the level of maximum density is closest to the ground (0 – 10 cm). The median of the altitudinal distribution of the biomass is a little less than 20 cm.

3. Vegetative biomass does not surpass 80 cm and more than 95% of it is below 50 cm in height. Between 80 and 140 cm only reproductive biomass was found.

4. In the late phase of annual development the biovolume of the dominant species, *Trachypogon vestitus,* was 33% of the total; those of the two subdominant species, *Leptocoryphium lanatum* and *Axonopus canescens,* 9% and 10% respectively. Together, the three most important species constituted 52% of the total biovolume, the 26 remaining species only 48%.

5. At the end of the rainy season, when the development of the total aerial biomass is at its maximum, the green biovolume was 35% of the total, the reproductive biovolume 10%, while the standing dead biomass was already 55% of the total. That is, the reproductive biomass diminishes steadily after the reproductive peak of *Trachy-*

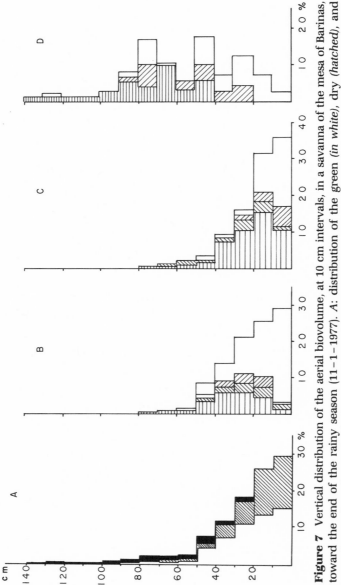

**Figure 7** Vertical distribution of the aerial biovolume, at 10 cm intervals, in a savanna of the mesa of Barinas, toward the end of the rainy season (11–1–1977). *A*: distribution of the green *(in white)*, dry *(hatched)*, and reproductive biovolume *(in black)*; *B*: distribution of the vegetative green biovolume; *C*: distribution of the dry biovolume; *D*: distribution of the reproductive biovolume. In *B*, *C*, and *D*, *Trachypogon vestitus* is marked by *horizontal hatching; Leptochoryphium lanatum by diagonal hatching bottom left to top right; Axonopus canescens by diagonal hatching top left to bottom right; other species shown in white.*

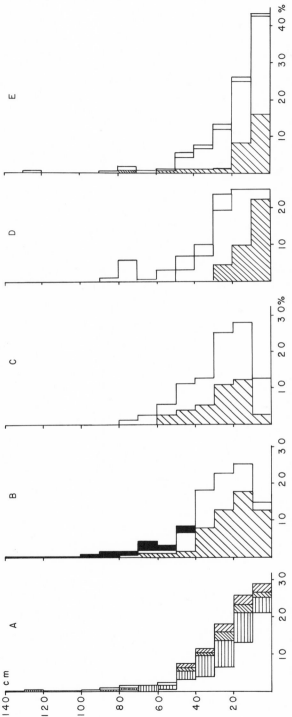

**Figure 8** Vertical distribution of the aerial herbaceous biovolume, at 10 cm intervals, in a savanna of the mesa of Barinas toward the end of the rainy season (11–1–77), by species. *A*: total biovolume; *B*: biovolume of *Trachypogon vestitus*; *C*: biovolume of *Leptocoryphium lanatum*; *D*: biovolume of *Axonopus canescens*; *E*: biovolume of the other species. (Dry is marked by *diagonal hatching*; green by *white*; reproductive by *black*.)

*pogon vestitus* in August-September and virtually disappears toward the end of the dry season. Green (living) biovolume is in a process of diminution and continues to diminish, while the volume of dead biomass, which already constitutes the major portion of the total annual volume and consequently an even greater part of the aerial biomass, continues to increase during the dry season until it forms almost all of the total above-ground biomass toward the start of a new yearly cycle.

6. The vertical distribution of the green and dry biomass is much the same. The living biomass decreases gradually and continually between 0 and 50 cm; its maximum value, 30%, is found between 0–10 cm, while only 10% is found between 40 and 50 cm. It then decreases abruptly and ends entirely at 80 cm. The dead biomass also has its maximum at the lowest level, where 36% of it is found, a value that decreases to 32% between 10 and 20 cm, and decreases more steeply after that until it disappears at 80 cm. That is, the dead biomass is concentrated in the two lower levels, so that its median is found below 15 cm. Since the green biomass is a little less concentrated in the lower levels, the median for the total vegetative biomass is at 18 cm above the ground.

7. The reproductive volume is more irregularly distributed between the ground and 140 cm in height, with maximum values between 40 and 50 cm and 70 and 80 cm. This irregularity reflects the fact that every species flowers at a different height.

8. The main part of the green biovolume is represented by the nondominant species (59%), even though they constitute only 48% of the total biovolume. Hence these species have more living tissue than the dominant ones. *Trachypogon vestitus*, which takes up 33% of the total biovolume, accounts for only 21% of the living biovolume; *Leptocoryphium lanatum* accounts for only 10% of the living biomass; and *Axonopus canescens*, for 8% — that is, it has a greater proportion of dead than living biomass, while the opposite is true for *L. lanatum*.

9. The other 26 species form 34% of the dry biovolume; *Trachypogon vestitus*, 47%; and *Leptocoryphium lanatum* and *Axonopus canescens*, 14% and 11% respectively.

10. The reproductive parts of *Trachypogon* are found between 40 and 140 cm, especially between 60 and 70 cm; in *Axonopus canes-*

*cens* the inflorescences are found between 20 and 90 cm with a maximum between 70 and 80 cm.

11. *Trachypogon vestitus* has a vertical distribution with a maximum between 10 and 20 cm (25% of its total biovolume), decreasing gradually up to 40 cm, and then more abruptly up to 80 cm for the vegetative portion and 140 cm for the reproductive portion. The biomass of this species also diminishes below 10 cm, so that the 0–10 cm layer contains only 15% of its total biomass. Of the total volume, 83% is dead tissue. Most of the living biomass (56%) is found between 20 and 40 cm, with 28% of the living biomass in each of the two levels 20–30 and 30–40 cm.

12. Most of the biovolume of *Leptochoryphium lanatum* is found between 10 and 20 cm, with the 20–30 cm level following; altogether 54% of the biovolume of this species is found on these two levels. In contrast to *Trachypogon vestitus* the greater proportion of the living biomass is found in the lower level.

13. The biovolume of *Axonopus canescens* is principally and evenly distributed among the lower three levels (0–30 cm), where 74% of its total living biovolume is found. Above this level the biovolume diminishes abruptly.

14. The nondominant species are found principally in the 0–10 cm level (43% of the total); after that their volume diminishes proportionally with height. The living and dead biomass of these species shows a vertical distribution very similar to that of the total biovolume.

The values obtained for total aerial biomass from the five harvested quadrats of 1 m² each were 633 ± 82 g/m². Of this, 22.5% was living biomass, 76.5% was standing dead biomass, and 1% was reproductive biomass. In comparing the biomass with the biovolume data, it is evident that the percentage of dead biomass is much greater than the corresponding biovolume, and the opposite is true for living biomass in relation to its biovolume. The distribution of the total biomass by species gave the following results: *Trachypogon vestitus*, 50.5%; *Leptochoryphium lanatus*, 9.4%; *Axonopus canescens*, 2.7%; and all the other species combined, 37.4%. Additionally, the green area index (GAI) obtained was 3.98 m²/m² and for the total surface (green and dead) the LAI was 7.6 m²/m².

In Japan scientists have analyzed grasslands and other secondary herbaceous communites (Monsi and Saki, 1953; Iwaki et al., 1964) utilizing the technique of the stratified harvest. They built "productive structure diagrams" of the vegetation (Monsi, 1968), in which the photosynthesizing and nonphotosynthesizing tissues are represented as a function of height. It was found that the nonassimilating biomass of the herbaceous communities diminished gradually and continuously, starting with the level closest to the ground, while the assimilating biomass increased with height until it reached a peak at 20 cm below the maximum height of the canopy of the community. In secondary grasslands of the subtropical areas of India, dominated by *Sehina* and *Heteropogon*, Shankar et al. (1973) also found a continuous altitudinal variation in the total biomass, as well as in the biomass of the two dominant grasses considered separately.

In summary, the results of the quantitative analysis of the vertical distribution of the biomass do not show any vertical stratification; they show only that the biomass gradually decreases with height. In the case at hand more than 90% of the biomass was found below 50 cm, although probably at the time of the blooming peak of the dominant species a greater, but still minor, proportion of the total biomass would have been above 50 cm. However, each of the dominant species, as well as the rest of the herbaceous flora, was distributed vertically in a slightly different manner, which shows that there is a vertical apportionment of space — a vertical distribution of the ecological niches. Finally, the annual cycles of the various species, as far as the proportion of living to dead biomass is concerned, were not synchronized with species such as *Trachypogon vestitus*, which at the time of the harvest had more than half of their tissues dried up, while others, such as *Leptocoryphium lanatum*, still had over 60% of living tissue in their biovolume.

These findings apply to the savanna at the end of the rainy season, but they are also applicable to other communities of the temperate and subtropical region.

## Stratification of the underground biomass of the herbaceous layer

For all the herbaceous plants of the savanna, the following is a valid generalization: in all communities underground organs are concentrated in the uppermost layers of the soil and quickly thin out with depth until they totally disappear at about 200 cm below

ground. Within this general framework, however, there are notable differences between the various savanna ecosystems in regard
to the concentration of biomass in the upper layers of the soil. These
differences are primarily due to the physical characteristics of the
soil. Figure 9 shows the distribution of the underground biomass of
the herbaceous stratum in four savanna ecosystems, at the time of
maximum development of the aerial biomass, that is, exactly at the
end of the rainy season. These data were obtained (Sarmiento and
Vera, 1978) by separating all the vegetal material in a vertical soil
column, at 20 cm intervals, from the surface to 2 m. The material
was sieved and decanted in the laboratory, then dried in an oven
and weighed. The fact that the sampling was done at the end of the
rainy season does not insure that the hypogeous biomass had its
maximum development at this time.

The figure shows that the bulk of the biomass is concentrated in

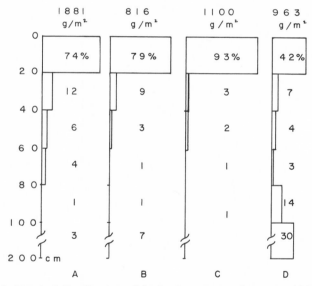

**Figure 9**  Vertical distribution of the herbaceous underground biomass at
20 cm intervals, in four savannas of the llanos of Barinas, at the end of the
rainy season (11–1–1973). A: *Leptocoryphium lanatum-Elyonurus adustus*
savanna (Boconoito); B: *Paspalum plicatulum-Axonopus purpusii* savanna
(Garza); C: *Sorghastrum parviflorum* savanna (Jaboncillo); D: *Axonopus
purpusii-Leptocoryphium lanatum* savanna (Barinas). The upper numbers
correspond to the total underground biomass at sampling time. (Data from
Sarmiento and Vera, 1977).

the upper 20 cm, and more than 75% of all the hypogeous biomass is above 60 cm. Below 1 m there are only traces of root biomass. Because these data will be discussed again in the chapter on primary production, here we only wish to point out the close relation between the distribution of the underground organs and the physical characteristics of the soil profile. For example, in the savanna of Jaboncillo, where there is a very compact and impermeable horizon at 40 cm below the soil surface, more than 90% of the underground biomass is concentrated in the first 20 cm. In the savannas of Boconoito, where the soil has better drainage and no major impediments for the development of the root system, the herbaceous plants are better able to extend their root systems into the lower soil layers, and the concentration of biomass diminishes more gradually with depth. The ecosystems of Barinas and Garza represent an intermediate situation, where the seasonal water table that can be as high as 130 or 75 cm respectively imposes limitations on the root development.

Figure 10 shows, in 10 cm intervals up to 60 cm depth, the results obtained from the analysis of the below-ground biomass and of the distribution of the aerial biovolume in the savanna in the Q4 terrace in the llanos of Barinas. At 60 cm a level of coarse pebbles prevents further root penetration. The results agree with those presented before, particularly in the concentrations of the biomass of savanna ecosystems, both above and below the surface.

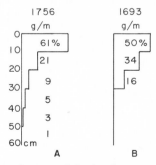

**Figure 10** Vertical distribution of the herbaceous underground biomass at 10 cm intervals, in two *Trachypogon vestitus* savannas in the mesa of Barinas, at the end of the rainy season (11 – 1 – 1977). The numbers indicate percentages of the underground biomass found at each level. *A*: savanna over an oxisol with pebbles at 60 cm; *B*: savanna over an oxisol with pebbles at 30 cm. The topmost numbers are the values of the total underground biomass at sampling time.

## The woody strata

The physiognomic-ecological definition of savanna need not be restricted to one with a herbaceous cover (grassy savanna); a savanna may also have woody strata of different degrees of height and cover. The nomenclature of the different kinds of woody savannas will be covered in the discussion of the horizontal structure. For now only the characteristics of the vertical structure of the woody strata and their contribution to the architecture of the ecosystem will be considered.

The woody species of the true savanna in northern South America are generally characterized by low stature. None of the species in these ecosystems grow above 8 m, and they commonly vary in height from less than 2 m to 6 or 7 m. In terms of Raunkaier's life-form classification, they would be classed as nano- or microphanerophytes.

There are many woody species in the savanna whose maximum height at maturity does not surpass that of the herbaceous stratum (50–80 cm). Hence these species, although morphologically different from herbaceous species, must be regarded as belonging both physiognomically and ecologically to the herbaceous stratum. Among them are species such as *Byrsonima verbascifolia* and *Psidium guinense* that never grow over 80 cm, as well as *Casearia sylvestris*, *Byrsonima crassifolia* and *Curatella americana*, which permanently maintain this "dwarf tree" morphology in many communities. This phenomenon is even more common in the cerrados, where there grow a large number of species that are considered to be "underground trees" (Rachid, 1947).

Sarmiento and Monasterio (1983) in their review of life forms in tropical savannas emphasize the differences between perennial species with entirely seasonal above-ground parts (trees), those with woody underground organs (the half-woody species), and the perennial herbs, with nonwoody rhizomes or bulbs. The half-woody species constitute one of the most peculiar life forms of tropical savannas. Some of them, the permanent geoxyles, conserve this habit under all circumstances and are therefore an integral part of the herb cover from a structural point of view. For the traumatic geoxyles, the half-woody adaptation results from external pressures, and they can revert to their normal tree habit when circumstances are favorable. These species belong to both structural vegetational layers: the herbaceous one when they are half-woody, and the tree layer when they outgrow the grass matrix.

**Figure 11** Three examples of wooded savannas with varying density and heights of trees. *Top:* open savanna in a mesa to the southeast of Calabozo (State of Guarico). Here *Bowdichia virgilioides* reaches up to 8 m in height, while *Byrsonima crassifolia* and *Curatella americana* barely reach 4 m. *Bottom:* savanna in the mesa of the state of Guarico with low woody species, primarily *Byrsonima crassifolia*, which determine its physiognomy. *Opposite:* a tree savanna in the south of the state of Apure with the dominant tree, *Caraipa llanorum*, reaching 10 m in height. Even though the trees are tall and may form small groves, this does not prevent the development of a herbaceous stratum with a cover close to 100%.

True members of the woody strata are those species that surpass the height level of the herbaceous cover and thereby give the savanna the aspect of a tree-savanna or a shrub-savanna (further on we will discuss whether the more common woody species of the American savannas should be classified as shrubs or trees).

The degree of cover of the woody strata varies in relation with the type of community that is being considered, from a few individuals per hectare (less than 1% of cover) to several thousand individuals per hectare. In the last case there would be a clearly identifiable canopy, with maximum values of cover of 60% to 70% (fig. 11c). But only in Brazil does one find forests formed by savanna trees of the magnitude of the sclerophyllous forest called the *cerradao*.

It is difficult to determine whether the woody elements form more than one layer in natural savanna communities. Simple observations show that within the range of heights of 1 to 8 m there are in most formations trees of all heights, and it would be arbitrary to distinguish any kind of strata. Only in some cases could the history of a stand have determined a discontinuity in the development of the woody species, so that strata would be indentifiable. In some circumstances strata may be determined on a floristic basis, given that there are species that attain different heights.

Quantitative data from Ataroff (1976) for two parcels of woody savannas in the piedmont of the llanos of Barinas (fig. 12) show that the two dominant species—*Curatella americana* and *Byrsonima*

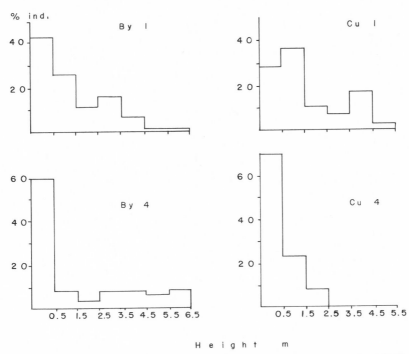

**Figure 12**  Vertical distribution of two tree species, *Byrsonima crassifolia* and *Curatella americana,* in two closed savannas (1 and 4) of the piedmont of Barinas. Each histogram was constructed on the basis of 100 randomly sampled individuals. (Based on data from Ataroff, 1976.)

*crassifolia* — sampled by measuring 100 randomly selected plants, exhibit a continuous height class variation or very irregular histograms. It is therefore difficult to determine levels or strata of maximum concentraion of the biomass, since the large number of small trees compensates for the greater biomass of the few larger trees. Eiten (1972) came to the same conclusion regarding the lack of stratification in the cerrado, and so did Cesar and Menaut (1974) for the woody species of the savannas of Lamto on the Ivory Coast.

## Distribution of the underground biomass of the woody species

There are no quantitative studies of the amount of below-ground biomass of woody species nor of its vertical distribution for any neotropical savanna, so the following discussion will of necessity be purely descriptive. Indeed, information regarding the distribution of the roots of woody species in different kinds of savannas is so fragmentary as to make any generalization very difficult. Scientists studying other tropical ecosystems have usually paid much greater attention to the above-ground than the below-ground component of the biomass. Only in herbaceous extratropical communities has the below-ground biomass been studied in detail, with the hope of understanding the functioning of the whole ecosystem and the dynamics of the component species. This work has been done in the North American prairies, thanks to the pioneering efforts of Weaver and his school, and in the Russian steppes and grasslands where Rabotnov and other ecologists worked. Their analyses of the hypogeous component of species and ecosystems have made it possible to reach a deeper understanding of the dynamics of the whole.

When studying the root development of savanna trees, it is necessary to keep in mind that even though each species has a genetically determined type of growth and root morphology, their expression varies tremendously with habitat, especially in spatial relationship, the degree of branching, and the maximum depth that roots can reach. Consequently one cannot deal with underground biomass without taking into consideration the characteristics of the soil.

The first methodical observations of the underground organs of the plants of the American savannas were published by Warming (1892). He pointed out the presence of very conspicuous underground organs in numerous woody and herbaceous species from Lagoa Santa, Brazil. Warming described the first cases of "underground trunks" in typical *campestres* species such as *Byrsonima verbascifolia* or *Casearia sylvestris,* among others. Warming also points out that the existence of prominent lignified organs immediately below the surface of the ground often makes it difficult to decide whether a species is an herb, a shrub, or a tree. According to Warming the presence of these organs is related to fire. Rawitscher and his colleagues (1943) and Rawitscher and Rachid (1946) also observed underground trunks and xylopods in the majority of the woody species of the cerrado of Emas in central Brazil. They noted that the roots of these woody species reached ten or more meters deep in these soils, and in some cases even reached the water table situated at eighteen meters. These authors point out that the development of the underground organs of these species is notably greater than that of the aerial ones, which are often reduced to leaves and inflorescences.

Van Donselaar – Ten Bokkel Huinink (1966) describes in detail the morphology and the type of underground development of numerous woody and herbaceous plants of different habitats from the savannas of the north of Surinam. This research shows how the horizontal and vertical distribution of the roots of *Curatella americana* varies principally in function of the depth of the water table. In savannas where the level of the water table comes close to the surface during the rainy season, this species acquires the form of a small shrub with a totally superficial root system with a dense crown located between 10 and 25 cm deep and lateral roots that extend horizontally up to 6 m from the base of the tree. In other savannas where the water table is situated at greater depths the individuals of *Curatella americana* not only become more tree-like in aspect, but the horizontal root system is found at greater depths, up to 40 cm, and it is notably extended with a radius of up to 27 m around the tree. In addition the main root acquires greater importance, and in well drained soils it penetrates several meters deep. Other species of the savanna have similar morphological responses to changes in the water relations of the soil.

Foldats and Rutkis (1965; 1969; 1975) have studied the development of the root system of the principal tree species of the Venezuelan savannas growing under different habitat conditions. They find that *Curatella americana*, *Byrsonima crassifolia*, and *Bowdichia virgilioides* have lateral roots that are very extended at depth levels of 20 to 50 cm, but they also develop a deep main root, unless there is a superficial water table, or rocky or hard lateritic layers that are difficult to penetrate.

In Africa Menaut (1971) studied Guinean and Ivory Coast savannas and noted that the principal "shrubs" of these ecosystems, such as *Cussonia barteri*, *Crossopterix febrifuga*, *Piliostigma thonninguii*, and *Anona senegalensis*, have thick vertical xylopods and an extensive horizontal root system similar to that found in the American savannas.

All these observations yield preliminary conclusions that may give some clues regarding the main strategies of savanna trees. In the first place it appears to be a general fact that the trees of the savannas have a very extensive root system. In this sense they are similar to typical woody species from arid regions. Secondly, the below-ground biomass can be divided into a dense system of horizontal roots, growing at depths of from 20 to 50 cm, and a deep vertical system. The combination of these two types of underground organs allows these species an exhaustive harvesting of the soil resources. Thirdly, there appears to be a vertical division of the edaphic resources between the two principal life forms of the savanna: perennial grasses and sclerophyllous trees. The grasses through their superficial root system tend to exploit the upper edaphic layers, and their below-ground biomass diminishes very abruptly below certain depths. The trees on the other hand have the bulk of their roots directly below the area of main concentration of the grass roots, and reach soil layers several times deeper than the herbaceous plants. Fourthly, the trees of the savannas seem to have a biomass allocation strategy of their aerial and underground tissues that is closer to a herbaceous than to a woody type. That is, while in general trees have a low root-to-shoot ratio (R/S), developing principally above ground, herbs in general show an evolutionary tendency to develop a greater portion of their biomass below ground, and consequently to have a higher R/S ratio. (By roots are meant all underground organs; most of the underground biomass may correspond to modified shoots, such as rhizomes, bulbs, and so on.) Since most savanna trees have a large underground biomass component, especially some of the "underground trees" like *Andira humilis* or *Anacardium pumilum*, their R/S ratios approximate or surpass that of most herbs and they may be thought of as "functional herbs." In the chapter on phenological rhythms we take up again the different behavior of grasses and woody species in relation to the temporal partition of the environmental resources.

# Horizontal structure

Following Godron (1968), horizontal structure is here defined as the physiognomic heterogeneity expressed in the horizontal plane that exists in a formation and can be detected by simple observation. The elements that contribute to this structure repeat themselves according to some determined distribution pattern. In savanna ecosystems the existence of trees and shrubs that form a discontinuous layer implies two different structural elements: one that is formed by the continuous herbaceous layer, the other formed by the herbaceous layer covered at least in part by the woody elements (fig. 13). This difference is independent of any changes that the trees may cause in the herbaceous layer or in other ecological conditions and attributes (microclimates, niche diversification, and so on) in each of the elements of the horizontal structure. A good example of the ecological changes in the horizontal structure induced by isolated trees was reported in the savannas of Belize (Kellman, 1979). In this open-tree savanna the soil below the trees showed preferential enrichment of nutrients in some cases approaching or exceeding the levels found in nearby rain forest soils. The capture of precipitation by the tree crowns is the major reason for the formation of these mineral enriched microsites where litter and herbage weight increase above the amounts found in the treeless spaces of the savanna. Obviously the importance of this phenomenon for the dynamics of the ecosystem does not need to be emphasized. Once the trees have been recognized as an integral part of the spatial structure, their precise pattern of distribution may be elaborated through the use of known methods for the identification and analysis of patterns and scales of heterogeneity (pattern analysis, Greig–Smith, 1964; Kershaw, 1973).

At a more detailed level of analysis, in many savannas it is possible to identify a horizontal structure within the herbaceous layer itself. It results from the dominant life form, the perennial bunch grass. In fact, the predominance of this life form constitutes in itself an ecological niche and a special microhabitat that differs from the one found in the free spaces between the grasses (fig. 14). The distribution of other species is influenced by this micromosaic of two habitats. Figure 14, depicting the savanna in the llanos of Barinas, shows that the spaces occupied by the tussocks of two species of grasses are different from the relatively open spaces between tussocks. In their analysis of the horizonal differentiation of this savanna, Trompiz and Silva (1982) found that the differences between these microsites, a question of centimeters, were statistically significant in terms of soil surface temperature, as well as in terms of amounts of calcium, magnesium, and phosphorus in the topsoil. Likewise, the frequency of grass seedlings was greater in protected microsites next to established tussocks. The differentiation also applies to the below-ground levels. Table 1 presents some data on the underground biomass in a seasonal savanna showing the large variation caused by the the horizontal microstructure of the herbaceous layer.

**Figure 13** The presence of isolated trees in the savanna can create a special microhabitat, as in the *top* picture showing low shrubs grouped around an isolated tree, or no change, as in the *bottom* picture showing a very open herbaceous stratum.

**Figure 14**   This savanna of the llanos of Barinas was photographed at the beginning of the rainy season. The herbaceous stratum has not yet reached its full development, permitting a view of the horizontal structure of tussocks separated by open spaces.

**Table 1.** Vertical distribution of underground biomass in a seasonal savanna in the space between tillers and below tillers of the dominant species: *Trachypogon vestitus, Leptocoryphium lanatum,* and *Axonopus canescens.* (Each measurement is the mean of 5 samples.)

| Horizon (cm) | Intertiller (g/m²) | Tillers (g/m²) |
|---|---|---|
| 0–10 | 223.6 | 1680.2 |
| 11–20 | 165.0 | 780.2 |
| 21–30 | 151.8 | 289.6 |
| 31–40 | 19.2 | 66.0 |
| 41–50 | 12.2 | 34.8 |
| 51–60 | 3.0 | 23.4 |
| 0–60 | 574.8 ± 65[a] | 2874.2 ± 828[a] |

[a] ± standard error.

In some cases the structure of the herbaceous layer is only a reflection of the heterogeneity of the environment predating the development of the vegetation. This happens for example with the *lombrizales* and the *zurales* (earthworm casts 10–20 cm in height, 5–10 cm in diameter) of many marshy savannas (fig. 15). But more commonly it is the vegetation that ultimately determines the differentiation of the habitat, and where there already is some heterogeneity, the vegetation tends to reinforce it.

The exact distribution of the trees throughout the herbaceous cover may be determined by various environmental factors. For example, trees may grow over termite nests, where there is better soil and more nutrients. Also common, but harder to explain, is the grouping of trees in areas where more or less disintegrated lateritic layers are found near the surface. However, environmental heterogeneity is more often the result and not the cause of the existence of woody species. In other words, it is the development of the vegetation that transforms a homogeneous environment into a heteroge-

**Figure 15** Grassy savanna of *Sorghastrum parviflorum* in a lowland of the llanos of Barinas, a few days after burning. The microrelief is produced by *zurales* or *lombrizales,* small mounds of biological origin 10–15 cm in height that create their own microhabitats or become structural elements with regard to water, air, and soil structure.

neous one, and the object of research is to determine what causes the patterns of distribution of the woody species.

The contribution of the woody trees to the structure of the savanna, whether in terms of cover or of density, and independently of the pattern of distribution of the trees within the community, determines to a large extent the physiognomy of the savanna. The whole physiognomic classification of savanna types is thus based on the woody species. Before discussing these classifications, however, it should be noted that tree density and cover are continuous variables; and that the physiognomic nomenclature of the savannas is neither uniform nor general, not even within one linguistic area of tropical America, much less in four different ones.

Since there is a continuum of variation of the density, height, and cover of the woody species (see Goodland, 1971, for a quantitative study), the delimitation of physiognomic types is fairly arbitrary. I have chosen a classification that is based on the most frequent physiognomic types found in Venezuela. Table 2 defines and distinguishes the different physiognomic units that were established following the criteria of a previous work (Sarmiento and Monasterio, 1971). The table also gives the most probable equivalents of the terms in use in Brazil (Eiten, 1972), the French terms used in the Ivory Coast (Cesar and Menaut, 1974), and the nomenclature proposed by English-speaking ecologists who work in East Africa (Pratt et al., 1966).

As here defined, the grass savanna has no woody stratum above the grassy one. The savanna may be either pure grassland or grassland with dwarf shrubs or trees no bigger than the grassy cover, and therefore only distinguishable physiognomically after a fire (fig. 16). The limit defining open and closed savannas has been set at 2 percent of the total cover of woody species, since at that value of cover the trees obstruct the vision in all directions, giving the observer the feeling of a "closed" horizon. After the cover of trees reaches 15 percent and movement through the terrain becomes fairly difficult, the savanna acquires the physiognomy that in Brazil is called *cerrado* (closed), which it is in the strictest descriptive sense (Eiten, 1972). There is no equivalent term in English or Spanish, so we have adopted that of woody savanna or woodland.

When the trees reach eight meters in height, they compose formations that are ecologically different from all the preceding ones, being closer to forests than savannas. For cover values below

**Table 2.** Physiognomic types found in American tropical savannas. For the more common formations the Spanish, Portuguese, and French names are also given.

1. Savannas without woody species taller than the herbaceous stratum
   *Grass savanna, grassland*
   Sabana pastizal
   Campo limpo
   Savane herbacée

2. Savannas with low (less than 8 m) woody species forming a more or less open stratum
   a. Shrubs and/or trees isolated or in small groups; total cover of woody species less than 2%
      *Tree and shrub savanna*
      Sabana abierta
      Campo sujo
      Savane claire et très claire
   b. Shrubs and/or trees with a total cover between 2% and 15%
      *Savanna woodland, wooded grassland, bush savanna*
      Sabana cerrada
      Campo cerrado
      Savane claire
   c. Trees with a total cover higher than 15%
      *Woodland*
      Sabana boscosa
      Cerrado (*sensu strictu*)
      Savane dense et très dense

3. Savannas with tall (over 8 m) trees; continuous or dispersed
   a. Isolated trees, with a total cover below 2%
      *Tall tree savanna*
      Sabana abierta alta
   b. Trees with a total cover between 2% and 15%
      *Tall savanna woodland*
      Sabana cerrada alta
   c. Trees with a total cover between 15% and 30%
      *Tall wooded grassland*
      Sabana boscosa alta
   d. Trees with a total cover above 30%
      *Tall woodland*
      Bosque claro
      Cerradão
      Forêt claire, savane boisée

4. Savannas with tall trees in small groups
   *Park savanna*
   Sabana parqueada

5. Mosaic of savanna units and forests
   *Park*
   Parque

**Figure 16** Savanna with dwarf woody species in a mesa of the state of Anzoategui. The individual specimens of *Curatella americana* will not be visible from a distance once the dominant grass (*Trachypogon vestitus*) reaches its maximum aerial development.

2%, or between 2 and 15% we will use respectively the same terms of open and closed savanna with the qualification of "tall." When, as is very frequent in the central llanos of Venezuela, the tall trees are arranged in groups *(matas)*, the terrain will be named "parkland savanna." Finally, the open scleromorphic forest (called cerradao), typical of the region of the cerrado, is not found outside of Brazil, so it will keep its own local name.

# 3

# The seasonal rhythms of the savanna species

Very little research has been done on the phenology of the species of the American tropical savanna. Warming, who pioneered the analysis of the seasonal developmental rhythms of these species in his great and classical study on *Lagoa Santa* (1892), has had very few followers. That is because until recently the importance of the phenological outlook was undervalued in most ecosystem analyses. Only toward the end of the 1960s, when the International Biological Program started to promote its comprehensive programs, did the study of plant phenology become recognized as an important element in comprehending the internal dynamics of an ecosystem, especially in understanding the productive processes. Clearly it is useful to have a precise knowledge of the periods of development of the photosynthetic organs and of the successive stages in the formation of the reproductive structures of grasses and other species in the tropical savannas. This information will be of direct use in range management, and it can also clarify the adaptive behavior and the evolutionary strategies of the still poorly understood components of these ecosystems.

Long after Warming, the phenology of the American savannas was studied again by Van Donselaar–Ten Bokkel Huinink (1966) in Surinam and by Monasterio (1968) in Venezuela. But it is only recently that Monasterio and Sarmiento (1976) published the first systematic analysis of the seasonal rhythms of a group of species of the Venezuelan savannas and suggested possible ecological reasons for each phenological strategy. Ramia (1977, 1978) extended the phenological observations to the flooded savannas of Apure State, thereby completing the picture of the phenological behavior of the most common species of the Venezuelan savannas. The most recent research done on the phenology of all tropical savannas is the work of Sarmiento and Monasterio (1983). It has placed the seasonal rhythms of the species of Venezuelan and neotropical savannas within a global framework. We will use this work as the main reference here.

The phenological observations of the seasonal savannas of the central llanos were made at the Biological Station of the Llanos in Venezuela (see fig. 2). This site presents a mosaic of units of open savanna and closed parkland savanna. The herbaceous stratum alternates between a period of active development characterized by a vigorous growth of the dominant species and a period of relative dormancy during the rainless season, when the dominant species gradually dry out. Toward the end of the dry season the bulk of the aerial biomass of this stratum is still standing, although it is dry.

The dominant species of these savannas are grasses of the genera *Trachypogon*, *Axonopus*, and *Andropogon* and sedges of the genus *Bulbostylis*. There are also many herbs and sufrutices of the genera *Cassia*, *Desmodium*, *Galactia*, *Eriosema*, *Indigofera*, *Hyptis*, *Zornia*, and so on. The savannas can be considered as floristically quite rich communities. To be precise, the total flora of the 300 ha of the protected land of the Biological Station consists of about 200 species (Aristiguieta, 1966), while the mean number of species enumerated in 380 censuses of 4 m² quadrats was 15 (Monasterio and Sarmiento, 1968). The woody layer is an open stratum formed by a few species of low, perennial, and sclerophyllous trees. The three most important species are *Curatella americana*, *Byrsonima crassifolia*, and *Bowdichia virgilioides*. These savannas may be said to belong largely to the *Trachypogon/Curatella* association, with some isolated semideciduous woods.

The principal climatic data for the region are summarized in figure 17. Of all the elements of the local climate, the seasonal distribution of the rains undoubtedly plays the most important role in the seasonal cycles of the plant species — this is of course true for the entire savanna area of the llanos. The contrast between the periods of rain and drought is complete. During the rainy season there is an excess of water, while during the four months without rain the surface soil dries out completely (for details see the chapter on water balance). There may also be great fluctuations in the yearly precipitations, and extremely dry years can be expected

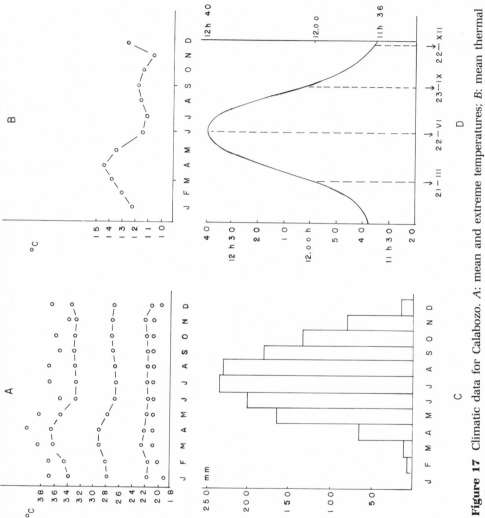

**Figure 17** Climatic data for Calabozo. *A*: mean and extreme temperatures; *B*: mean thermal oscillations; *C*: precipitation; *D*: photoperiod. Months are identified by initials. (Data according to Monasterio, 1970.)

**Table 3.** Phenological characteristics of species from the savanna of the Biological Station of the Llanos.

| Species (phenological group in parentheses) | Family | Frequency | Life form | Height (cm) | Leaf size | Periodicity | Perennating organs |
|---|---|---|---|---|---|---|---|
| **Herbaceous** | | | | | | | |
| *Aeschynomene brasiliana* DC. (5) | Leguminosae | Infreq. | G | 40 | Nano | Seasonal | Xylopod |
| *Aeschynomene hystrix* Poir. (4) | Leguminosae | Infreq. | Th | 10 | Nano | Seasonal | Seed |
| *Andropogon brevifolius* Sw. (3) | Gramineae | Freq. | Th | 30 | Nano | Seasonal | Seed |
| *Andropogon semiberbis* (Nees) Kunth (1) | Gramineae | Subdom. | H | 120 | Meso | Seasonal | Rhizome |
| *Aristida capillacea* Lam. (3) | Gramineae | Freq. | Th | 20 | Nano | Seasonal | Seed |
| *Axonopus canescens* (Trin) Pilger (1) | Gramineae | Domin. | H | 80 | Micro | Seasonal | Rhizome |
| *Axonopus chrysoblepharis* (Lag) Chase (1) | Gramineae | Subdom. | H | 80 | Micro | Seasonal | Rhizome |
| *Axonopus purpusii* (Mez) (1) | Gramineae | Domin. | H | 50 | Micro | Seasonal | Rhizome |
| *Borreria suaveolens* Mey (4) | Rubiaceae | Freq. | Th | 40 | Nano | Seasonal | Seed |
| *Bulbostylis capillaris* Kunth (1) | Cyperaceae | Subdom. | H | 40 | Micro | Seasonal | Rhizome |
| *Bulbostylis conifera* Kunth (1) | Cyperaceae | Subdom. | H | 30 | Micro | Seasonal | Rhizome |
| *Bulbostylis junciformis* (H.B.K.) Kunth (5) | Cyperaceae | Subdom. | H | 80 | Micro | Seasonal | Rhizome |
| *Diectomis fastigiata* (Swartz) H.B.K. (3) | Gramineae | Subdom. | Th | 80 | Micro | Seasonal | Seed |

| Species | Family | Frequency | Life form | Height | Size | Duration | Regeneration organ |
|---|---|---|---|---|---|---|---|
| *Digitaria fragilis* (Steud) Luces (3) | Gramineae | Infreq. | Th | 30 | Micro | Seasonal | Seed |
| *Eragrostis maypurensis* (H.B.K.) Steud. (3) | Gramineae | Freq. | Th | 15 | Nano | Seasonal | Seed |
| *Evolvulus sericeus* Sw. (6) | Convolvulaceae | Freq. | H | 15 | Nano | Perennial | Xylopod |
| *Galactia jussieana* H.B.K. (6) | Leguminosae | Freq. | H-Ch | 50 | Micro | Perennial | Xylopod |
| *Indigofera pascuorum* Benth (6) | Leguminosae | Freq. | H | 80 | Nano | Perennial | Xylopod |
| *Microchloa indica* (L.F.) Kuntze (3) | Gramineae | Infreq. | Th | 10 | Nano | Seasonal | Seed |
| *Pectis carthusianorum* Less. (4) | Compositae | Freq. | Th | 10 | Nano | Seasonal | Seed |
| *Pectis ciliaris* L. (4) | Compositae | Infreq. | Th | 10 | Nano | Seasonal | Seed |
| *Polycarpaea corymbosa* (L.) Lam. (3) | Caryophyllaceae | Infreq. | Th | 10 | Lepto | Seasonal | Seed |
| *Tephrosia tenella* Gray (4) | Leguminosae | Infreq. | Th | 10 | Lepto | Seasonal | Seed |
| *Trachypogon plumosus* (H.B.K.) Nees (1) | Gramineae | Domin. | H | 120 | Meso | Seasonal | Rhizome |
| *Trachypogon vestitus* Anders (1) | Gramineae | Domin. | H | 120 | Meso | Seasonal | Rhizome |
| *Zornia reticulata* Sw. (5) | Leguminosae | Freq. | G | 40 | Nano | Seasonal | Xylopod |
| **Arboreal** | | | | | | | |
| *Bowdichia virgilioides* H.B.K. (2) | Leguminosae | Domin. | MiPh | 600 | Micro | Perennial | Woody organ |
| *Byrsonima crassifolia* (L.) H.B.K. (1) | Malpighiaceae | Domin. | MiPh | 600 | Meso | Perennial | Woody organ |
| *Curatella americana* (L.) (2) | Dilleniaceae | Domin. | MiPh | 600 | Meso | Perennial | Woody organ |

with sufficient frequency to constitute an important ecological factor to be reckoned with. As for the changes in day length and its variation during the year, the tropical zone offers an interesting contrast to the northern latitudes in that the hottest (April) do not coincide with the longest days (June). On the other hand, the coldest time of the year almost coincides with the shortest days (December–January). It is evident that the separation of two environmental factors of great importance for the regulation of the phenological rhythms (thermo- and photoperiod) offers more species the possibility to develop a thermo- or photorhythm according to whether they are responsive to one or the other of these environmental stimuli.

The species whose seasonal rhythms were analyzed are tabulated in table 3, which also lists their most important ecological and morphological features, such as the stratum where they are found, their relative abundance, their life form, the size of the leaves, the

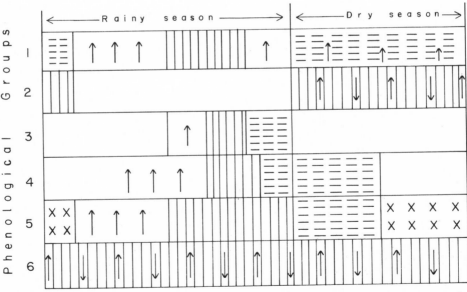

**Figure 18** The phenorhythms of the species of the savannas of Calabozo. Each numbered row corresponds to a phenological group. *Up* arrow indicates leafing; *down* arrow indicates leaf fall; *horizontal hatching* corresponds to the period of drying of the aerial biomass; *vertical hatching* corresponds to the reproductive period; *crosses* indicate vegetative growth; *white blocks* indicate the periods when the species exist only as seeds in the soil. (Modified from Monasterio and Sarmiento, 1976.)

periodicity of the foliage, the type of perennating organs, and the phenological group to which they belong. Of the thirty species studied, three are evergreen microphanerophytes, fourteen are hemicryptophytes, and the thirteen remaining ones are annual species, some of them very small and short-lived.

Taking into account the vegetative and reproductive annual cycles of the species, we have grouped them in six phenological types, each of which is characterized by a determined seasonal pattern (fig. 18). Each of these six behavioral patterns will be examined in turn, without forgetting that the species included in each group do not all follow a totally homogeneous type of behavior. However, even though species in each group may exhibit differences in detail regarding the beginning or the end of some annual rhythm, there is a notable similarity in their total rhythmicity.

# Group 1. Perennial species with an annual semidormant phase

Almost all the dominant species of the herbaceous stratum of the savannas belong to this group. It is composed of Gramineae or Cyperaceae whose only perennating organ is the rhizome, from which the entire aerial and subterranean part of the plant is formed each year.

The phenophase of vigorous growth begins with the start of the rainy season (end of April to May), or after a fire if the savanna has been burned toward the end of the dry season, and produces the photosynthetic plant organs. In a few months (between the end of June and August) the vegetative growth reaches its annual peak, after which, in the middle of the rainy season, it slows down very abruptly in direct relation to the beginning of the reproductive phases. Within a few weeks inflorescences develop, blooming begins, and the normal sequence of stages follows until seeds have matured. After seed dispersal (from September to early November) there may be a short period when all the green biomass remains stable, until the point is reached when increasing drought imposes a limit to the maintenance of the shoots. As the ground dries out (starting in December), the shoots yellow and die; although new shoots are still being formed from the rhizomes, they don't develop beyond the stage of a few leaves. Finally these shoots dry out and die too, together with almost all the aerial biomass. The plant enters the phase of semidormancy that is characterized by the increasing reduction of the green biomass together with the formation of a few shoots that never grow beyond the early stages of development.

Figure 19 provides a more quantitative view of the behavior of a species with this annual phenological rhythm. It shows that vegetative growth is concentrated in a relatively brief period and describes how the green, dead, and total biomass varies during the year. Despite a decline of the green biomass, the total standing biomass remains relatively constant until the beginning of a new cycle, or until it is totally consumed by fire at the end of the dry season. This constancy of the standing biomass indicates that the rate of decomposition of the aerial biomass is approximately equivalent to the rate of assimilation, while the rate of drying of the aerial biomass is much greater than the rate of growth during the period between the peak of vegetative growth and the end of the annual cycle.

Other important species of Gramineae and Cyperaceae, which are found in the savannas of the western llanos more often than in

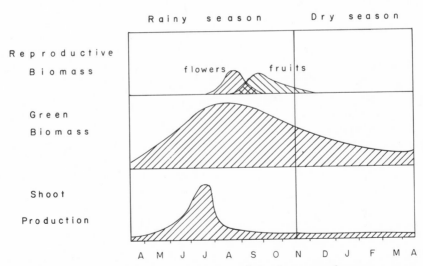

**Figure 19** Semiquantitative phenogram of *Trachypogon plumosus*, a perennial grass of phenological group 1. Because the vertical scale is only relative, no numerical values are given. The *lower* curve represents the production of shoots; the *central* curve, the changes in green biomass; the *upper* one, the changes in reproductive biomass (flowers and fruits). (From Monasterio and Sarmiento, 1976.)

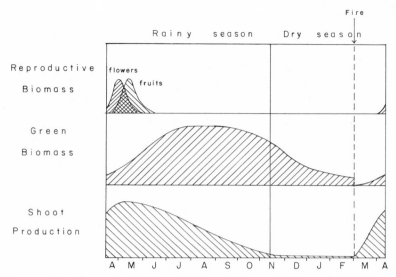

**Figure 20** Phenogram charting the development of *Leptocoryphium la-natum*, a perennial grass of phenological group 1 in a grassy savanna in the mesa of Barinas, subjected to fires from the beginning of March. See figure 19 above.

the central llanos, show in their annual cycle the same alternation of an active phase with a semidormant one, but their phenological behavior during the growth phase is different. Thus, for example, *Leptocoryphium lanatum*, *Elyonurus adustus*, and *Bulbostylis para-doxa* start their growth phase by developing the reproductive structures first, and only later growing their new photosynthetic organs (fig. 20). This pattern of precocious blooming at the beginning of the wet season, or even before if the savanna is burned, necessarily implies the existence of subterranean storage organs with enough reserves to allow completing the reproductive cycle without first developing the photosynthetic tissues, except for whatever the inflorescences themselves may be able to contribute in assimilates. The possible advantages of this strategy compared to the more normal pattern for this group are not clear. One advantage of this phenorhythm of precocious blooming may be that it allows the dispersed seeds at the beginning of the rainy season to avail themselves immediately of conditions of humidity favorable for germination.

In view of these behavioral differences, the species in group 1 can be divided into two subgroups: the first comprises species that bloom after the development of the vegetative body of the plant (1A); the second comprises species that bloom early, either before or simultaneously with growth of the photosynthetic structures.

When the savanna is not burned, the standing biomass of the dominant species of this phenological group decomposes slowly over a period of years during the rainy seasons. The total amount of epigeous matter hardly varies from one season to the next over a period of several years, unless there are big differences in rainfall (see the chapter on productivity). What changes is the state of this biomass. After the start of the rainy season there is an explosive increase in the green biomass as well as an increase in the rate of decomposition of the dry biomass left from the previous year; later on in the cycle the standing dry biomass increases until it totally dominates at the end of the cycle. In any case, with or without fire, the green biomass is renewed every year; only the underground rhizomes persist from one cycle to the next and have a chance to increase their biomass.

The phenological pattern that has been described for this group of species is particularly well suited to the ecological conditions of the seasonal savanna. It is therefore conceivable that this pattern has played a role in the ecological success of these species, which have become the dominant elements of the ecosystem. Their behavior appears to be a direct adaptive response to the stress brought about by seasonal drought. During the favorable months of the year these species maximize photosynthetic assimilation and conclude the whole reproductive cycle; afterward, they enter into a phase of gradual decline that reduces the activities of the plant to a minimum at a time when the environmental stress becomes very great. However, this phase of partial inactivity does not seem to be genetically determined; on the contrary, it appears to be a flexible adaptation that allows these genotypes to survive in this environment. Proof of the lack of genetic program is that activity does not cease during the unfavorable season and the plants continue to put out shoots.

Most of the biomass of the herbaceous layer of the savanna belongs to species that follow this phenological strategy; in addition, a good proportion of their total biomass grows below ground, in the upper layers of the soil. Examination of the vertical structure of an open savanna in the mesa of Barinas (western llanos, savanna of *Trachypogon-Leptocoryphium*, ecologically and physiognomically similar to that of the Biological Station) disclosed that the first 20 cm of the soil contained 74% of the hypogeous matter and 69% of the total biomass at that moment. This soil stratum constitutes, without a doubt, the most seasonal part of the ecosystem as far as humidity conditions are concerned, since it follows strictly the changes in rainfall. The intensive exploitation of the water resources at this level allows the plants to initiate their growth as soon as the first rains of the wet season fall; likewise, and for the same reason, when the rains end the plants will dry, following the drying rhythm of the soil. Another important element of this strategy is that the slow decomposition of the aerial standing biomass allows an economic use of the mineral elements, thanks to a pattern of reallocation within the plants as the shoots dry out (see the chapter on mineral cycles). This behavior is of course very important for species that are suffering a severe nutrient limitation.

The competitive chances of these dominant species may also be heightened by their permanent presence — they form a more or less closed cover during the whole year. This herbaceous cover, whether green or dry, maintains an effective occupation of space, favoring the ecological success of its species by preventing access to others.

The frequent natural and man-made fires have a special impact on this group of species. They occur mostly toward the end of the dry season and consume almost completely the standing epigeous biomass, which is not only all but dry but has also lost a large portion of its nutrients through translocation. At the same time, the seed bank that was produced several months before is lying in the soil and is relatively protected from the destructive action of the flames. The negative effects of a fire thus appear to be minimal. However, fire does induce the beginning of a new annual growth cycle, and if the burning takes place considerably before the rainy season it will undoubtedly have an unfavorable effect by triggering growth processes that will later not be able to continue owing to the lack of appropriate moisture conditions.

Not only do all the perennial grasses of the nonflooded savannas show the same phenological behavior, but also a relatively numerous set of these species coexist within a given community, several among them being codominants. Within a phenological pattern that is similar in its basic response to the seasonal water stress, the various grass species differ in their seasons of growth and reproduction. The very precocious species that initiate growth as soon as the rainy season starts, or even before then (just after a fire) are followed by early species that reach their vegetative and reproductive peak two to three months after the rainy season has started; after them come the intermediate species that peak well into the wet season; and last are the late species that concentrate their growth and reproduction at the end of the wet season or beginning of the dry one. In effect there is a temporal partition of the niche of similar life form species that coexist in the same ecosystem. Even though one or another form may dominate depending on total amount of rainfall and length of the season, as a general rule these four forms coexist in the herbaceous stratum. The explanation for this coexistence may be found in the uncertainty of some regulating factors, such as the timing of burning or the beginning of the rainy season. This uncertainty would favor the maintenance of the phenological diversity. Silva and Ataroff (1982) and Canales and Silva (1982) have been analyzing the effect of this temporal niche partition on the reproductive strategies of these grasses, and quantifying the growing rates and seasonal patterns.

# Group 2. Evergreen trees with seasonal growth

This group encompasses all the tree species of this savanna, since they all show the same phenological pattern: the new leaves and flowers are produced during the dry season, while during the rainy season morphogenetic activity ceases altogether.

Leaves start to fall at the beginning of the dry season in a long and gradual process that continues throughout the dry period; at the same time that old leaves are falling, new ones are growing. Because these opposite processes are proceeding simultaneously, the total surface and biomass of the leaves decrease during this period, but the plants always have some foliage. At the start of the rainy period, or shortly thereafter, growth ceases altogether, no new leaves are formed, and even the leaves that were developing remain arrested at the stage that they had attained at the time. Throughout the rainy season the foliage remains green, and its chlorophyll content per surface unit reaches a peak (Medina et al., 1969), although toward the end of this season many leaves acquire a brownish tint that indicates the beginning of senescence.

The reproductive process also takes place almost entirely during the dry season, but the ripening of the seeds and their dispersal may continue for a time after the beginning of the rainy season. Germination of seeds appears to take place very seldom in these savannas; toward the end of the rainy season only a very few seedlings of *Curatella americana* could be seen. Figure 21 shows the phenological rhythm of this species; it is interesting to note that the aerial growth reaches a maximum precisely in the middle of the dry season, simultaneously with blooming and fruiting, and coinciding with the yearly minimum in the quantity of green matter.

At first glance this phenological strategy may appear surprising. In fact these trees carry on their principal activities — the growth of branches, formation and expansion of leaves, flowering and fruiting — during the least favorable climatic period, when there is practically no input of rainwater. What may be the advantages of this behavior? To be sure, leaf fall does produce a notable diminution of the assimilatory surface and of total transpiration of each individual; it takes place as drought increases and thereby contributes to the maintenance of the water balance. On the other hand, as

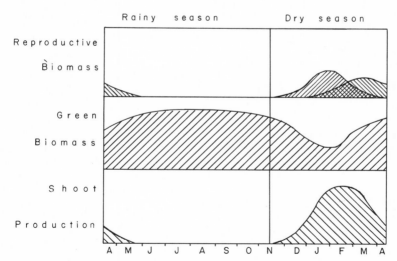

**Figure 21**  Phenogram of *Curatella americana,* a woody species belonging to phenological group 2. The system of representation is the same as that in figure 19. (From Monasterio and Sarmiento, 1976.)

these trees undoubtedly have the capacity of growing and developing during the dry season, they must have a reliable source of water, even though the soil water potential could be relatively low. The root system of these species (see chapter 2) is very extensive, so that a median-sized individual (5 to 6 m) may have a system of lateral roots that extends over more than 20 m from the foot of the tree, at a medium depth of 20 to 50 cm. From these lateral roots grow vertical ones that frequently reach depths of 6 m or more. Because there is an abundance of water during the rainy season, it is reasonable to assume that the deeper layers of the soil may retain enough water the whole year around to allow the growth of deep-rooted species. In certain cases the roots might reach the permanent water table. However, this is not necessary for the survival of the trees; on the contrary, too high a water table has an adverse effect on the woody species in these savannas.

The trees use the whole of the rainy season for maximum photosynthetic assimilation, but they leave the renewal of the assimilatory organs to the dry, least favorable period. In contrast to the grasses that enter their phase of semidormancy as the upper layers of the soil dry out, the trees continue growing, since they exploit the resources of deep soil layers. Underground, too, there is the same kind of spatial partitioning of the ecological niche between the two dominant biological forms in the ecosystem that was observed in the aerial space.

The fires that take place during the dry season consume some of the senescing leaves that still remain on the trees. They can also damage new buds and leaves, especially on young individuals, since older plants are largely above the reach of the flames. Fire, if it takes place every year, sculptures in this manner the external form of the trees but does not destroy them. The effect of fire on the soil seed bank is not yet known, but when the savannas burn, the seeds are either not formed or are still maturing on the trees. However, the survival of the woody species in the savannas appears to depend more on vegetative propagation than on sexual reproduction, and the former is, without a doubt, stimulated by fire. Indeed, after a fire the development of shoots and inflorescences is explosive, growth rates are exceptionally high, and all of the assimilatory apparatus is reconstituted in a few weeks. This is a general phenomenon, valid not only for the three species of trees studied in Calabozo, but also for all the species of the Venezuelan savannas.

# Group 3. The ephemeral species

While the two preceding groups are formed by the dominant species of the ecosystem, the next four include species of the herbaceous stratum that are of little quantitative importance in the totality of the vegetation. Nevertheless, their existence points to certain phenological alternatives that have permitted other species to adapt to the savanna environment.

Among the flora of the Biological Station eighteen annual species have a short life cycle. Of these, eight were chosen for study: seven grasses and one of the Caryophyllaceae *(Polycarpaea corymbosa)*. Most of these annual species are small in size and together they make up the low herbaceous stratum of the savanna; but whatever their size all are ephemeral, in the sense that they complete their cycle in less than two months.

The annual activity of this group does not start until the middle of the rainy season (July–August), which makes it the last phenological group to become active. During the whole first part of the wet season (that is, from April to June) the plants exist only as seeds — more or less protected — in the ground. The phenological dynamics may be divided into three stages that follow each other quite rapidly. The first stage is germination, followed by a fast vegetative development and immediately by blooming. These phenophases succeed each other so quickly that it is said that these plants *germinate blooming,* that is, they reach their reproductive maturity almost instantly, or, to put it another way, they bloom when they are still seedlings. At the same time distinctions between vegetative and reproductive organs are blurred, as all the shoots are destined to become reproductive, so that inflorescences are morphologically not distinguishable from the vegetative axis; the two are merely different phenophases of a single structure.

In the second stage the seeds mature while the vegetative apparatus starts to senesce and dry. In the third stage the mature seeds remain on the dry plant until they are dispersed. The seed bank can remain in the soil eight to ten months until the start of a new cycle. This cycle applies to the whole set of populations of the eight species of ephemerals that were analyzed; individuals among them may complete their cycle in a still shorter time, of four to six weeks.

The ephemeral species have very superficial roots, so that they only use the resources of a very thin layer of the soil and are thus very sensitive to climatic variations. Their vegetative apparatus is reduced to a bare minimum, with many small leaves that are short-lived. These plants represent the extreme range of species with an *r* strategy, in that they tend to maximize the production of a large number of small seeds. Because the energy allocated to this function has to be exceptionally high, the entire individual plant may even be considered as a reproductive structure.

Once their short life cycle has been completed, the seeds of these species remain in the soil for a long time. Their chances of germination during the next favorable season depend in some measure on the presence of an adequate microenvironment, since they grow in empty spaces or where there is not too much interference from other vegetation. Such niches are not found every year. The plants of this group prosper in those areas where cover is low (due to unfavorable habitat conditions such as sandy soils, rocky sites, and so on), or where there has been some sort of disturbance. The exceptions to this pattern are *Digitaria fragilis* and *Andropogon brevifolius*, two annual grasses that grow in the shadow of the high herbaceous layer, which can therefore be singled out as a subgroup of shade-loving ephemerals in contrast to the other sun-loving species.

The ephemerals, like all annuals, have low competitive ability. Each species germinates massively under adequate conditions, forming colonies that are almost monospecific. From the moment a seed germinates, only an accident can prevent it from completing its cycle until the production of seeds; that is, the environmental filter operates almost exclusively at the stage of germination. Because the entire life cycle of the ephemerals takes place during the wettest season of the year, these species escape both fire and drought. The fact that seeds germinate as late as three months after the start of the rainy season suggests either that germination is controlled by a water-soluble germination inhibitor, or that the seeds are under photoperiodic control, germinating only after the longest days of June – July. This last hypothesis is reinforced by the nature of the open habitat where these plants grow and by the fact that their seeds germinate only in the sun.

The strategy of the ephemerals is characteristic of desert plants. Its occurrence in the seasonal, wet environment of these savannas demands an ecological explanation. These species do not prosper in the truly arid tropical formations (*espinares* and *cardonales*); they are only found in certain habitats that have been mentioned (very dry or disturbed) within the area of the savannas. Such a special type of adaptation suggests to us a certain kind of relictual flora that originated in steppe-like seasonal communites and might have prospered in these tropical regions during more arid climatic phases.

# Group 4. Annual species with a long cycle

If we exclude the ephemerals, the annual flora of the Biological Station consists of 54 species. Of these, approximately half (28) belong specifically to the grassland-savanna element, while the other half (26) grow only among the trees or in the borderland between the savanna and the interior forests. We followed the life cycle of five long-lived annual species of the savanna flora; all of them occupy a very modest place within the total biomass of the herbaceous stratum of the savanna, although they may occasionally take over a site that has been disturbed.

The activity cycle of the annual species takes place during six to seven months (see fig. 18), almost entirely within the rainy season. Germination occurs one or two months after the beginning of the rainy season, followed by a vegetative phase that lasts two or three months, until the beginning of the blooming period. The rest of the cycle is dominated by reproductive activities. Blooming can last up to four months, with uninterrupted formation of new flowers; during this phase vegetative growth may continue, so that leaves, flowers, and fruits are produced concurrently. When the wet season comes to an end these plants decline quite fast, although the fruits may remain on the dry stalks for a long time. Finally, the gradual period of seed dispersal having come to an end, the seeds may remain in the ground for five or six months until the beginning of a new cycle.

The phenological strategy of the annual species is entirely directed toward the maximization of the reproductive effort. They behave therefore as r species (Gadgil and Solbrig, 1972), since most of their life cycle is spent in producing seeds while their vegetative development is greatly restricted.

From an ecological point of view, many of the annual species with a long cycle may be described as weeds of disturbed sites. By contrast, the five that we have analyzed are typical elements of the natural savannas and can maintain themselves there indefinitely.

To explain the existence in this ecosystem of the phenological strategy of annual species with a long cycle we must take into consideration three factors: the existence of drier habitats, where the herbaceous cover is more open and the annuals acquire more importance; the annual variability in rainfall, which may cause drier years with a truly semi-arid climate; and the proved occurrence of drier climatic phases in the past, so that these species could be relicts of a flora that was once more abundant.

# Group 5. Perennial species with a seasonal rest period

The principal phenological characteristic of this group of herbaceous species is a rather long period in their annual cycle during which they are completely inactive. They are hemicryptophytes or geophytes whose aerial organs are renewed every year, while their underground organs remain in a latent state until the beginning of another wet cycle. Their aerial biomass is totally annual, and in this respect their phenological behavior is like that of the annual species; however, in contrast to true annuals, this group has perennial underground organs. This survival strategy, so common in cold climates, is here also an adaptive response to a season that is unfavorable for growth because of the annual drought. In the savannas of Calabozo only about ten species exhibit this strategy. Among them the following three have been analyzed: *Bulbostylis junciformis*, *Aeschynomene brasiliana*, and *Zornia reticulata*.

*Bulbostylis junciformis* is a compact member of the Cyperaceae that has the peculiarity of being the only species among the dominants in the herbaceous stratum with an annual phase of total inactivity. Its active period lasts from five to eight months, starting with the wet season and extending into the beginning of the dry season. Growth starts with the first rains; two months later the inflorescences start to elongate and blooming takes place between July and the beginning of August. Immediately afterwards the plants start showing signs of senescence, so that as the seeds are maturing the aerial part of the plant is drying until the plant is totally dry but still standing.

The other two species in this phenological group are legumes that are found quite frequently in these savannas but constitute only a minor element of the total herbaceous stratum. Their perennating organs are thick xylopods that serve as storage organs of organic matter and minerals. In these two species vegetative growth is maintained throughout the entire rainy season, accompanied by blooming from August on. But in November, even before the beginning of the dry period, leaves start to fall more or less rapidly, and the whole plant dries out.

When the wet season is unusually long, all three species may experience some vegetative growth after blooming, but this last growth cycle does not lead to blooming; the branches dry out before the cycle is completed.

Most of the geophytes of this savanna (Iridaceae, Araceae, Orchidaceae) show a similar seasonal behavior but have a longer latency period. They start their active cycle with the coming of the rains and complete it in three or four months, so that by the middle of the rainy season the plants have disappeared completely. That is, the latent phase of their hypogeal life is up to nine months long. However, this phenological group exhibits two different types of blooming behavior, just as did group 1 (herbaceous species with an annual phase of semidormancy). One subgroup (5A) includes species that bloom late and only after a prolonged phase of vegetative growth, like the three species analyzed above. The species in the second subgroup (5B) begin their cycle with flowering, and only afterward complete the development of the assimilatory apparatus. In general the precocious species have a shorter cycle than the late blooming ones and may be considered true "ephemeral perennials" because of their ephemeral aerial structures. A typical early bloomer is *Curculigo scorzoneraefolia* (Amaryllidaceae), which flowers in April just a few days after the fires. Its leaves persist for only a few weeks and are gone by June.

By examining how species avoid the unfavorable season, we can compare the phenological strategy of the perennials and the annuals. The former have permanently established underground organs, thus enabling them to maintain more easily a place in the ecosystem in the face of competition from the other groups of perennials. They take advantage of accumulated underground reserves to grow very rapidly when conditions turn favorable, before the dominant grasses and Cyperaceae can develop their biomass completely and occupy the totality of the aerial space. In addition, the resting phenophase of the dormant perennials coincides with the dry season while their growth takes place in the wet season; these species are therefore unaffected by drought or fire and consequently are very well adapted to the savanna environment. Their only problem appears to be to complete their life cycle fast enough to minimize competition. Two conditions must be fulfilled to meet this challenge successfully: sufficient reserves must be accumulated to allow almost explosive growth when conditions become favorable; and there must be an external stimulus to cause these species to synchronize their development with the favorable season.

# Group 6. Species with continuous growth and blooming

This phenological group—of evergreen perennial species that grow throughout the year—is characterized by a behavior that is completely unlike that of the previous three groups. Three sample species of this group are: *Evolvulus sericeus, Indigofera pascuorum,* and *Galactia jussieana.* The three have a similar morphological organization, with a xylopod that serves as a permanent reserve organ with roots growing out of it that may be as much as one meter long, and a number of ramifications, each of them with a limited life span, which are formed throughout the year.

Of the three species, *Evolvulus sericeus* shows the least annual rhythmicity. This small subshrub produces shoots thoughout the year; they in turn continuously produce flowers. Each ramification, after emerging from the xylopod, develops by forming leaves and axillary flowering buds. After a few months the branch dies, having formed some twenty leaves and flowers. It is then replaced by another, and so on thoughout the year. Although the aerial development units are seasonal, the plant as such always has green leaves, is always growing, and continuously flowers and produces fruits.

The other two species of this group are also active throughout the year but manifest a certain rhythmicity in regard to the dry and wet seasons. These species do form new leaves during both seasons, but in the dry season they experience a partial loss of leaves that markedly reduces the total green biomass. But toward the end of the dry season the production of new leaves picks up, and soon the foliar surface characteristic of the wet season is restored. On the other hand blooming takes place at any time throughout the year. It is not synchronized within the population; instead, individual plants bloom at different times and there is always some specimen in bloom at any one time.

To be able to sustain this phenological strategy under such opposite conditions, the plants must have an adequate supply of water throughout the year, as well as enough reserves of energy to grow in the least favorable times of the year. In addition, to overcome the destructive effects of fire, these subshrubs must possess mechanisms that will allow them to renew growth afterwards. The morphoecological characteristics required to carry out this strategy may be: first, the existence of powerful underground organs that at any time constitute a large part of the total biomass of the plant; second, the permanent development of short-lived vegetative organs permitting the expenditure of energy and materials to occur in a series of consecutive short bursts, so that the destruction of any number of branches will not jeopardize the survival of the individual. The reproductive effort also takes place in a series of consecutive short steps. This appears to be a good general strategy for functioning in an environment that varies more than the sharply seasonal environment of the savanna. These species appear to possess a strategy to cope with a more extreme and unpredictable environment such as, for example, a semi-arid climate, where this opportunistic strategy would take advantage of favorable conditions at any time of the year. It could also be considered a typical strategy for living under stress, with a continual activity but at a low rate.

## Summary of phenological strategies in Venezuelan savannas

Study of the savanna species shows clearly the great range of possibilities in the behavior of the flora living in an environment of cyclical oscillations, especially of the marked seasonal variation in soil humidity. In the course of a year the herbaceous species (groups 1, 3, 4, and 5) alternate between a phase of minimal activity that always corresponds with the dry season and a phase of accelerated development and maximal activity that corresponds with the wet season. During the peak of activity photosynthetic assimilation is intense, all the photosynthetic apparatus is renewed, and reproduction takes place. By contrast, during the period of reduced activity there is either a phase of semirest when the green tissues are reduced to a minimum (group 1), or a phase of total dormancy or latency during which all aerial tissues die and only the perennial underground organs survive (group 5). As for the annual species of short or long cycle (groups 2 and 3) only the seed bank remains alive during the dry season to renew their presence as soon as favorable conditions return. Fire may induce the change from the latent to the active phase, but the possibility of normal growth thereafter depends on the first rains coming soon after the fire.

The phenological strategies of the species of the first group show certain characteristics that explain their adaptation to the seasonal environment of these savannas and also suggest the keys to their ecological success in the community; they form the morphological link between the structure and the functioning of the ecosystem. In the first place, the vegetative growth rhythms of these species are perfectly coupled to the periodicity of the rains. The plants decline during the period of water shortage and develop quickly the moment there is enough available water in the upper layers of the soil. This fast growth is possible because the perennial underground organs of these plants are able to harvest the upper layers of the soil as soon as they become wet. Furthermore, these plants preempt the space permanently, either when they are growing actively, or when there is a great deal of dead biomass left standing. For these reasons the herbaceous perennials dominate in their environment over the annual species and the species with a complete rest period. They have two of the characteristics that, according to Grime (1974, 1977), are common to species with competitive type strategies: maximum height in relation to concurrent species, and maximal production of leaf litter (in this case dead standing biomass).

The coupling of the vegetative phenophases with the seasonal rhythm of the rains also allows these plants to be flexible in responding to the vagaries of the environment. This flexibility can have a decisive effect on their persistence, given the great annual variations in the climate. The quantity or periodicity of the rains in any given year is rigorously reflected in the growth and subsequent decline of these species. Only fire or mowing can alter this synchronization by speeding up germination. The reproductive phases, on the other hand, apparently respond to more constant external signals, such as photoperiod; one set of species in subgroup 1A blooms after the longest days of the year, while the species with precocious blooming, such as *Leptocoryphium lana-*

*tum* and others (subgroup 1B), may be responding to a thermoperiodic stimulus before the beginning of the rains (they bloom when both air and soil temperatures are highest).

This pattern of growth is also adaptable to savanna fires, because when a fire takes place the epigeous biomass is more than 90 percent dry, and the live organs are well protected below the surface. Moreover, the fire undoubtedly has the effect of triggering a new growth phase, so that its occurrence at the right moment, that is, a few days before the arrival of the rains, favors the plant's development.

With regard to the balance of carbon and mineral nutrients, the herbaceous species have three characteristics of undoubted adaptive value in an ecosystem that is so poor in nutrients. In the first place, and in contrast to the woody species, these species do not maintain passive structures whose construction has required a great investment of energy and materials in short supply and which immobilize the available nutrients for a long time. Second, these plants recycle the mineral elements internally by relocating them to the below-ground organs when the aerial apparatus starts to dry out and before it is burned (see chapter 6 on the economy of nutrients). In the third place, the plants probably possess a high rate of photosynthetic assimilation when water balances are not limiting, since for the most part they are C4 species. On the basis of preliminary measurements made under laboratory conditions, Torres and Meinzer report rates of up to 20 mg $CO_2 \cdot dm^{-2} \cdot hr^{-1}$ for two savanna grasses *Trachypogon vestitus* and *Leptocoryphium lanatum*. These are rather high rates for plants growing on very poor soils.

The local populations of bunch grasses and Cyperaceae maintain themselves in the ecosystem by growth and vegetative reproduction, starting from a small number of individuals, by means of continual growth and fragmentation of the rhizomes. The competitive importance of sexual reproduction is thus reduced, and the selective filter against seedlings thereby loses some of its ecological significance. This reproductive pattern is in accordance with *K* type strategies common to species adapted to stable environments.

The annual plants (groups 3 and 4) also follow a phenological strategy adapted to a seasonal environment with a prolonged dry period. However, the proportion of these species to the total ecosystem biomass is quite small, despite their numerical importance in the local flora of the savanna (almost a fourth of the flora). The annual species are not capable of competing with the perennial grasses, but there are enough available niches to allow them to maintain a stable equilibrium within the system.

The perennial species that have a phase of total dormancy (group 5) share the existing resources with the dominant grasses through a temporal, rather than a spatial, division of these resources. They concentrate their maximum annual effort in the first months of the rainy season, completing their entire cycle before the total development of the dominant species. In contrast, both the semiwoody species and the trees of the savannas (groups 2 and 6) display certain phenological characteristics that are quite opposite to those of the herbaceous species. In the first place they are active throughout the year: the trees maximize photosynthetic assimilation during the wet season and dedicate the dry season to the renewal of their aerial organs and to reproduction; the subshrubs show an uninterrupted annual pattern of vegetative and reproductive activity. The strategy of these species of continual activity may be equated with what Grime (1974) calls "strategy of resistance to stress."

Two morphoecological characteristics help the woody species to overcome the less favorable season. The first is an extensive and deep root system that not only allows them to utilize water and mineral resources from a considerable volume of the soil and in particular from the deeper layer which the herbs cannot reach, but also functions as a reservoir of energy and nutrients during the unfavorable season. Secondly, even though these species do not regulate transpirational losses well, they reduce their cuticular transpiration to a minimum; also, because of their highly accentuated foliar scleromorphy, they can tolerate high levels of dessication without irreversible damage. In this manner they behave like plants that both evade and resist the drought, since they exhibit simultaneous control and tolerance of the transpirational losses.

Without a doubt, fire is one of the most powerful environmental filters in relation to the woody species. In the area of the Biological Station, where there has been fire protection of the savanna since 1961 (even though part of it suffered the effects of two accidental fires in 1968 and 1971), the density of trees has increased (San Jose and Farinas, 1971). The most destructive effect of fire is on the juvenile individuals that have not overcome a critical height threshold. Only exceptional fires reach the foliage of adult trees. While juveniles can be kept by fire from reaching that threshold for a long time, sooner or later the population overcomes that point and thereby reduces the destructive effect of fire to a minimum. That is, the important characteristics are the speed of growth in order to take advantage of the fire-free period and the capacity to resprout continuously following the damage produced by the flames.

In summary, examination of the seasonal rhythms in the savannas of the llanos reveals a diversity of phenological patterns that allows the species to overcome in one form or another the environmental restrictions. From a phenological point of view, the species are saturating the environment; that is, the flora is occupying, through the articulation of very elaborate strategies, all the possible phenological niches in the ecosystem.

# Phenology of the species in other tropical savannas and their comparison with the Venezuelan savannas

In a study of the form and periodicity of various species of the savannas of the north of Surinam, Van Donselaar-Ten Bokkel Huin-ink (1966) lists 177 species and indicates the period of the year when each one has green functional foliage. For this purpose the year is divided into four seasons: a long wet period (April–June); a long dry season (August–November); a short wet period (December–January); and a short dry season (February–March). In reality both dry seasons are periods of diminished precipitation but not of drought, since in this wet tropical area with 2000–2300 mm a year of rainfall only two months (September–October) have less than 100 mm (Van Donselaar 1966) and none less than 50 mm.

The author distinguishes eight phenological types constituting the savanna-like vegetation of the region, where characteristic savanna and open forest species are found on white sands and savanna species on latosolic or hydromorphic soils. They are:

1. Evergreen species with green functional foliage throughout the year. This group of 48 species includes all the trees and the shrubs, except for six that belong to the next group. Six herbaceous species from very wet white sands are also included here.

2. Species whose foliage decreases during the long dry season, including five woody species and thirteen herbaceous ones. Among

the woody species are *Curatella americana* and *Byrsonima crassifolia*, and among the herbaceous ones is *Paspalum plicatulum*, belonging respectively to groups 1 and 2 of the plants in the Venezuelan savannas.

3. Plants with total dormancy during the long dry period. This is a more numerous group and includes annuals, geophytes, and hemicryptophytes (*Trachypogon plumosus, Leptocoryphium lanatum, Axonopus pulcher*, and so on). We placed the hemicryptophytes in a different category of only partial dormancy, because although their foliage dries out they continue to produce new shoots in 90 percent of the cases.

4. Species that have functional foliage only during the long rainy season. All the species (eight in total) are annual, as *Aristida capillacea, Cyperus amabilis, Scleria micrococca*, and so on; these also exist in the llanos and belong to group 4 of annuals of long cycle.

5. Plants with green foliage at any time that conditions are favorable. Only a few annual species belong to this group, such as *Polygala adenophora* and *Polygala longicaulis*. They correspond to the ephemeral species in our classification, differing only in that they are opportunistic.

6. A single species that is active in the principal dry season and in the short humid period that follows.

7. Two species of geophytes with functional foliage only in the short wet and short dry seasons.

8. Two species of geophytes (*Scleria hirtella* and *Habenaria sp.*) that are active during the short wet season.

It is not always possible to translate the preceding groups into the terms of our classification, especially since only one phenophase has been considered in this study on the savannas of Surinam — the period of the year when the foliage is green — without taking into account the reproductive phenophases. Nevertheless it is evident that these species have several strategies in common with those of the Venezualan savannas, as for example the woody species that change their foliage in the dry season (group 2); the herbaceous species with a period of semidormancy (group 1); herbaceous species with a period of total rest (group 5); annuals of long cycle (group 4); and annuals of short cycle (group 3). The authors do not indicate any group of species with continuous growth and flowering (group 6), although such species do exist in the savannas of Surinam and probably exhibit there the same phenological strategy they show in Venezuela.

On the other hand, some of the phenorhythms from Surinam cannot be found among the Venezuelan species of known phenology, as, for example, the growth pattern of certain woody and herbaceous evergreen species that show no synchronization in the loss of their leaves. In Surinam these species grow in swamps and *morichales*. Nor do we find in Venezuela plants with the strategy of the annual species with an opportunistic short cycle; in Venezuela these are the ephemeral species with a short cycle at the beginning of the rainy season. The difference can probably be explained by the lack of drought in the north of Surinam, where these species do not need to limit themselves to a single favorable annual period. For the same reason the geophytes that have an active aerial life during the short wet period and some of the short dry periods are not represented in the Venezuelan savannas.

The phenorhythms of the flora of African savannas were described in a very thorough study of the secondary savannas of Nigeria (Hopkins, 1970), in which a quantitative analysis of the annual growth patterns of 17 species was made: 14 phanerophytes, 4 hemicryptophytes (three grasses and one sedge), and 3 geophytes. The general growth pattern is very similar in all the species despite certain minor variations, and therefore Hopkins maintains that it is impossible to differentiate or classify phenological types. All the species studied start their growth cycle very abruptly in February, after the final fires of the dry season. At the beginning growth is very intense (short flush growth) and is more prolonged for the herbaceous cover than for the trees, which continue growing, though with lesser intensity, only until July. The loss or drying out of foliage starts with the end of the rainy season and reaches its peak in December–January. That is, the patterns of growth and desiccation of the aerial biomass correspond very closely to the humidity and fire cycles. The most remarkable quantitative datum of Hopkins is that the maximum foliar growth rate of grasses and sedges may reach $12-19$ mm $\cdot$ day$^{-1}$ at the time of the growth flush.

One of the most exhaustive studies of several different types of tropical savanna ecosystems was performed at the Lamto Station in the Ivory Coast. The phenology of the herbaceous species of these ecosystems was analyzed by Cesar (1971), and that of the woody species by Menaut (1971). The savannas of Lamto are classified as Guinean or prewoody savannas of Occidental Africa. Their climatic conditions, although similar to the ones of the llanos in annual rainfall and temperature (1200 mm; 26°C), differ in that the dry season is much shorter; only two months (December and January) have rainfall below 50 mm · month$^{-1}$ (Lecordier, 1974).

Cesar's figures on the annual march of the green biomass in burned savannas and on blooming and fruiting of the dominant grasses, *Andropogon schirensis*, *Hyparrhenia chrysargyrea*, and *H. diplandra*, reveal a phenological behavior similar to that of the dominant species in the savannas of Calabozo (group 1A). These grasses increase their epigeous vegetative biomass more or less regularly from the moment of burning (middle of January) until they reach their maximum standing green biomass toward the end of the wet season (September to November), at which time they bloom and fruit. After that the green biomass decreases abruptly, and the cycle is repeated after the savanna has been burned once more. The drying out process varies among different dominant species. Some, such as *Loudetia simplex*, still appear relatively green at the time of the fire. These species, which Cesar describes as "late blooming, immediate emergence, fast growth, and partial drying," are similar to the dominant species of the savannas of Calabozo in that they follow the same phenological strategy as the perennial species with a period of semirest.

Other dominant grasses in the savannas of Lamto, such as *Imperata cilindrica* and *Brachiaria brachylopha*, have a similar cycle of development and senescence of aerial biomass but differ in their precocious blooming, which starts immediately after a fire. Such early blooming is shown by numerous other species that make a small contribution to the total biomass of the herbaceous stratum, like *Bulbostylis aphyllantoides* and *Cyperus schweinfurthianus*. This type of phenological strategy also corresponds exactly to that of certain dominant species of the savanna of the llanos characterized by early blooming: *Elyonurus adustus*, *Leptocoryphium lanatum*, *Bulbostylis paradoxa*, and others. The African species, like their American counterparts, begin a surprisingly fast foliar growth after the fire; Cesar calculates an average of 2–3 cm per day. These species resprout immediately after the fire, and in general they maintain themselves green for a longer period than other herbaceous species.

As far as the annual species are concerned, the only grass Cesar mentions is *Sorghastrum bipennatum,* an annual species with quite a long cycle of ten months. It germinates after the first rains, blooms at the end of the rainy season, and disappears very quickly thereafter. This annual can become dominant in closed and in woody savannas; for this reason, and also because of its long life cycle, it does not have any equivalent among the species of the Venezuelan savannas. Most of the remaining annual species (species of *Tephrosia, Indigofera, Aspilia,* and so on), also have a relatively long cycle, from April–May to December–January and are equivalent to the annual species of the savannas of the llanos (group 4).

At Lamto there are also some species with a phase of total dormancy during which the aerial tissues dry out completely, and only the underground perennial organs remain. Among these species we find *Borreria octodon* and *Vernonia guineensis,* of which the first species is late blooming and the second one is precocious. Their phenological strategy is similar to our group 5 of perennial plants with a rest period. This group also includes the many geophytes such as *Urginea indica (Liliacea bulbosa)* and *Curculigo pilosa,* that have perfect equivalents in the savannas of South America in species of Amaryllidaceae and Iridaceae such as *Curculigo scorzonearifolia* and *Cypella linearis.*

In addition, Cesar describes some species with very prolonged reproductive activity and almost continuous growth during the entire year, such as the geophytes *Aframomum latifolium,* and *Aneilema setiferum,* that could be compared to our group 6 of arhythmic perennials. However, the ones at Lamto disappear from the surface during the dry season and consequently fit rather in between our groups 5 and 6. Finally, the study mentions a single species of very short annual cycle: *Phyllanthus sublanatus,* an ephemeral that fulfills its life cycle in a few weeks and is therefore comparable to our group 2 of ephemeral species. It differs from species in that group by the surprising characteristic of being able to complete its life cycle at any time of the year when humidity conditions are favorable, displaying a maximal opportunistic strategy that has not been encountered in the American savannas. *Phyllanthus stipularis,* a small ephemeral of group 3 that grows in the savannas of Calabozo, is somewhat similar but has a fixed annual cycle.

The results of Cesar's study of the savannas of Lamto and their comparison with the results of work done in the Venezuelan savannas are summarized in figure 22, which describes the strategy groups classified by Cesar and their equivalents in our nomenclature. Four, possibly five of the phenological groups are common to both kinds of savannas, but two strategies are only found at one site: the ephemerals of short cycle that appear during the last part of the rainy season have no equivalent in Lamto, and opportunistic ephemerals are not represented in Venezuela. Nor are there perennials with uninterrupted activity in Lamto, although there are species with a very prolonged flowering period.

In his discussion of the phenological behavior of the trees of the savannas of Lamto, Menaut (1971) indicates phenorhythms that are essentially the same for the four most important species: *Crossopterix febrifuga, Bridelia ferruginea, Piliostigma thonningii,* and *Cussonia barteri,* as well as for six other less common species. The phenological pattern is the same: the trees lose their leaves starting

I     Precocious species of annual cycle (bloom at the beginning
      of the rainy season or immediately after a fire)

        *Long cycle* (perennials with a period of
        semidormancy)                     Group  1B

        *Short cycle* (perennials with a period of
        dormancy)                          Group  5B

II    Species with a late annual cycle (bloom in the middle or
      the end of the rainy season)

        *Immediate emergence* (vegetative development at
        the beginning of the rainy season)

            Annuals of long cycle             Group  4

            Perennials of long cycle         Group  1A

            Perennials of short cycle        Group  5A

        *Delayed emergence* (vegetative development in the
        middle or end of the rainy season)

            Annuals of short cycle          Group  3

            Perennials of long cycle         Group  1A

            Perennials of short cycle     (only in Lamto)

III   Species with continuous activity         Group  6

IV    Opportunist species (annuals of short cycle, develop at
      any time of the year)             (only in Lamto)

**Figure 22** Synoptical key of the principal phenological strategies of the species of the herbaceous stratum of American and African tropical savannas. (Data from Cesar, 1971, and Sarmiento.)

at the end of the rainy season, with maximum loss in January, which is amplified by the burning that takes place at that time. Formation of new foliage, which may start already before the fire, accelerates thereafter, so that in two months the foliage is totally developed. Blooming takes place simultaneously (March – April until July), followed by the maturation of the fruits. Only *Cussonia barteri* shows a behavior that is different in some aspects from the rest of the woody species. Its leaf formation continues throughout the year, and it can even lose all its foliage at any time if a drought takes place, and then produce a new crop of leaves. Its reproductive pattern too is aberrant in relation to the other species, since some individual trees fail to bloom every year; the ones that do start flowering in April, and the fruits ripen between June and December.

All the woody species of Lamto analyzed by Menaut may be classed as evergreen or brevideciduous, since they renew their foliage simultaneously with the fall of the leaves from the previous year, without any perceptible rest period. In this respect they clearly differ from the trees that are characteristic of the formations of the deciduous tropical forest (Monasterio and Sarmiento, 1976), showing instead a complete similarity with the woody species of the Venezuelan llanos, all of which are either evergreen or brevideciduous (the distinction between these categories depends on the period of time that the trees may be bare between the loss and renovation of their foliage). Also, according to Menaut, the behavior of these species is very flexible, varying with the pluviometric conditions of each particular year; when rainfall is above average, they behave like true evergreen species.

The most notable difference between the trees of the African and the Venezuelan savannas is that all the African species have a long period of fruit maturation. Flowering generally coincides with the start of the rainy season (April) and the period of fruit ripening and seed dispersal lasts until the end of the rainy season and even into the next dry season of December – January, so that during the period of burning some trees may still have fruits from the preceding cycle. The trees of the savannas of northern South America flower in the dry season or a little earlier, but in a period of two months the fruits mature and disperse their seeds. During the first months of the rainy season (May – June) it is unusual to find fruits on the trees, except for some dry structures from which the seeds have already been removed. In those cases where germination was

observed in Lamto it occurred from August on, from seeds produced during that same cycle. It is not clear whether this difference
in the time of seed maturation is related to to morphological characteristics, or whether it has any ecological significance.

Taking as variables the seasonality of the assimilatory structures,
the growing and flowering rhythms, and the length of the life cycle,
Sarmiento and Monasterio (1983) recognize fifteen phenological
groups in tropical savannas. A first distinction separates the species
with continuous carbon assimilation during the entire year from
those with seasonal assimilation, that is, those with a rest period.
Species with continuous carbon assimilation are further grouped
into those with continuous growth and those with seasonal growth.
In addition, each of these two groups as well as those with seasonal
carbon assimilation can be further divided on the basis of their
blooming times. According to this scheme, the perennial grasses
are divided according to their blooming times in the various groups
of the category of species with continuous growth and assimilation,
while the majority of the woody species experience continuous
assimilation during the year but discontinuous growth. Only a few
have more than one period of growth, while some species (deciduous) have seasonal assimilation. The remaining herbaceous species (other than grasses) and the annual and perennial subshrubs
constitute the phenological group with seasonal assimilation. Only
one group among these life forms has both continuous growth and
assimilation. This manner of grouping shows the diversity of possible phenological strategies in the savanna and their relation to the
different life forms, and the diversity of ecological conditions that
are found in these ecosytems.

# 4

# The productive processes

The productive processes of neotropical savannas have been studied fully only in Venezuela; more limited studies have been carried out also in Costa Rica. Productive processes are those ecological processes that participate in the balance of the organic matter of an ecosystem: photosynthetic assimilation, reallocation of the biomass among the different functional compartments, temporary accumulation of biomass in certain organs, mortality of the standing biomass, and its final decomposition, whether it takes place epigeously, hypogeously, or in the litter layer. The net result of these activities, which normally occur simultaneously, is the more or less continuous variation of the quantity of vegetable matter along the annual cycle, and in certain cases also a net increment in the biomass of the aerial or subterranean perennial organs.

The turnover rates of carbon in the vegetation largely determine the contribution and the modalities of permanence of this element in the soil, thereby regulating its biochemical cycle. The carbon cycle of the savanna ecosytems will be discussed in the chapter on nutrient economy; for the moment we will only consider the changes that occur within the assimilation component of the ecosystem.

The results obtained for the productive processes in several savanna ecosystems of Venezuela will be compared with some tropical savannas, with grasslands of the temperate zone, and with various forest ecosystems in order to point out the similarities and the differences among them. This study will not take into account the secondary production or the biomass consumed by herbivores, but will always refer to the net primary production of the system as the balance between the net primary production of the vegetation and the amount consumed by native herbivores (Odum, 1971).

Although the available information is both fragmentary and insufficient, the author has preferred to risk presenting a number of hypotheses and to sketch a few functional models of the productive processes of the savannas in the hope of stimulating new investigations. But before discussing these models it will be useful to explain their conceptual and methodological basis and to test their validity in one of the better known savannas of Venezuela.

## Biomass and production in Venezuelan savannas

The variations in the biomass throughout the annual cycle and the primary production of the Venezuelan savannas have only recently begun to be studied in a systematic fashion. In 1962 Blydenstein published the first data about biomass and production as part of a regional study of the savannas of *Trachypogon* in the central llanos, but these preliminary studies were not continued and extended to other ecosystems until the middle of the decade of the 1970s.

Extant studies cover three regions of the llanos (see fig. 2): the region of Calabozo, especially research at the Biological Station, where the *Trachypogon* savanna was studied (San José and Medina, 1975, 1976, 1977; Medina et al., 1977; San José et al., 1982); the region of Mantecal and Hato El Frío, in the state of Apure, where the so-called savannas of *banco, bajío* and *estero* were studied (Escobar and González Jiménez, 1975, 1979; Bulla et al., 1977, 1980; Entrena et al., 1977); and the savannas of the llanos of Barinas, where studies were conducted on four communities along a humidity gradient (Sarmiento and Vera, 1978). In addition, there are isolated data on the biomass of the savannas of *Trachypogon* in the llanos of Monagas (Espinoza, 1969) and a study of the flooding savannas of *Paspalum fasciculatum* in the south of the state of Guarico (Escobar and Medina, 1977; Escobar, 1977).

Some of the results known to date come from preliminary studies of an almost exploratory nature, or work that was undertaken with a different objective where productivity data were only a secondary aspect of the investigation. Consequently, despite the recent efforts of several working groups, productivity is still among one of the least known aspects of the Venezuelan savannas and vast regions of savannas and many formations remain totally unknown. The studies that exist provide acceptable information only on aerial herbaceous production, but the picture regarding underground production continues to be very confused, and there are only fragmentary data regarding the behavior and the biomass of woody species.

Nevertheless, the Venezuelan data are almost the only available ones for all the neotropical area and are therefore of exceptional interest. We will briefly review the major studies, pointing out in each case major contributions as well as arguable points, in order to try to produce general conclusions and to point out the most promising research lines.

In the terminology used here biomass is the total quantity of standing vegetable matter at a given moment, regardless of its conditions. It can be either epigeous or hypogeous. The epigeous biomass may be either live and assimilatory or standing dead and dry (straw or mulch in a grassland). Moreover, the total epigeous vegetable matter is the biomass plus the litter, dead matter separated from the plant and lying on the ground. The total hypogeous organic matter of the system is the total underground biomass plus humus. Unless the contrary is specified, vegetable matter will include its organic and mineral constituents, since it is often an arbitrary decision whether the mineral elements in the plant are or are not united to the organic matter.

Blydenstein (1962) studied the herbaceous aerial biomass in the *Trachypogon* savannas of the Biological Station of the llanos in Calabozo, in conditions that excluded grazing and under several burning regimes. The average precipitation in this region is around 1300 mm a year, with four dry months (December to March). Three plots were burned, one in December (1960) and the others in January and March (1961). By the end of August and September 1961, that is, in the middle of the rainy period and in the middle of the flowering season, the dominant grasses reached 533.8, 228.9, and 230.4 grams $\cdot$ m$^{-2}$ of dry weight respectively. On the other hand, a parcel that had not been burned since 1959, at the beginning of the growing season (April) had an aerial biomass of 439.4 g $\cdot$ m$^{-2}$, reached by August a biomass of 680.6 g $\cdot$ m$^{-2}$.

These measurements show that the minimum aerial herbaceous production of the burned parcels in eight months would be between 228.9 g $\cdot$ m$^{-2}$ and 533.8 g $\cdot$ m$^{-2}$; and in five months, for the unburned control, of 241.4 g $\cdot$ m$^{-2}$. Blydenstein does not give the error in his measurements, based on the harvest of eight quadrats of 1 m² each at the beginning and the end of the analyzed intervals, and consequently one cannot estimate the significance of the differences between the burned quadrat and the control. In any case the values for the three quadrats are very similar (228.9, 230.4, and 241.4 g · m⁻²), but very different for the quadrat burned in December, whose production was almost double the others (533.8 g · m⁻²). This last parcel corresponds to a savanna of *Trachypogon vestitus*, while the other three belong to the savanna of *Trachypogon montufari*. However, Blydenstein does not ascribe the difference in amount to the difference in the dominant species, nor does he give any other explanation for the anomalous value.

In the same work he presents data about the root biomass in two parcels of *T. montufari* and one of *T. vestitus*, obtained at an unspecified time of the year, through the separation of the hypogeous biomass of 10 soil blocks, 20 × 20 cm and 10 cm deep. The first two parcels yielded a radical biomass of 161.5 and 243.2 g · m⁻², while 292 g · m⁻² were collected for the plot of *Trachypogon vestitus*. It is clear that the radical biomass of this last savanna appears to be superior to that of the other community.

These figures only hint at the order of magnitude of the aerial and subterranean biomass of these savannas. Their fragmentary character does not permit a reliable evaluation of their annual primary production, since in addition to the sampling problems and the lack of any data on sampling error, the increment in the herbaceous biomass between September and March is unknown, and so are the rates of decomposition throughout the year. Moreover, the year the measurements were made (1961) was drier than normal, with 1136 mm of rainfall that did not begin until the end of May, so that a smaller than normal productivity could be expected.

In the following year Blydenstein again compared the total herbaceous aerial biomass as it was in August in parcels of the same savannas that had not been burned for four, two, and one year. The most interesting conclusions from these additional data are as follows:

The biomass accumulated in the different parcels in the March – August 1962 period (251 to 404 g · m⁻²) was significantly greater on the average, and for all measurements and treatments, than the mean of the previous year. The mean increment of the epigeous biomass was 35% higher. This figure can be correlated with precipitation since the season was 22% rainier than the previous year.

The biomass accumulated in the burned parcels was greater than that produced in the unburned ones, but the difference was not important and most likely it is not statistically significant.

The total aerial herbaceous biomass in the parcels that were not burned during two successive years reached 850 g $\cdot$ m$^{-2}$; while in those parcels that were protected during four consecutive years it reached 1000 g $\cdot$ m$^{-2}$. These figures demonstrate how very slowly the dry standing vegetable matter decomposes in the course of several years.

Twelve years after Blydenstein's 1963 work, new data about production of the savannas of *Trachypogon* were elaborated by San José and Medina, working at the Biological Station of Calabozo (1975, 1976). These authors followed the annual cycle of the living aerial biomass and the leaf area index (LAI) in two parcels: one had been burned on 28 December 1968, while the other had not been burned since 31 December 1964. The three dominant species (*Trachypogon montufari*, *Trachypogon plumosus*, and *Axonopus canescens*), which together account for 99% of the aerial biomass, were tabulated separately.

The maximum live aerial biomass and the maximum value for the leaf area index were obtained at the end of August 1969 when they reached 325 g $\cdot$ m$^{-2}$ (biomass) and 4.19 m$^2$ $\cdot$ m$^{-2}$ (LAI) in the unburned quadrat, and 415 g $\cdot$ m$^{-2}$ and 4.95 m$^2$ $\cdot$ m$^{-2}$ in the burned one, respectively. Since the total aerial biomass during the annual cycle or the rate of decomposition have not been calculated, there is no way to estimate precisely the total value of the aerial production beyond the peaks of maximum green biomass. Nevertheless, this study shows for the first time in an unequivocal manner that the increment of green biomass was significantly greater (about 30%) in the parcel burned halfway into the dry season, than in the protected one. Most of the increment occurred in the biomass of the two subdominant species (*T. plumosus* and *A. canescens*). But this does not prove, as some scientists assumed, that there is more total production; this cannot be determined on the basis of the data that were published.

A comparison of the maximum values of green biomass derived from this study with those obtained by Blydenstein shows that the figures for 1969 are higher. The magnitude can be correlated with precipitation, which in that year was much above the mean (1879 mm).

San José and Medina (1977) present additional data that facilitate an estimate of the savanna's aerial production. So, for instance, the burned parcel showed an annual peak of total epigeous biomass of 731 g · m$^{-2}$, which was computed at the same time as the already mentioned peak of maximum green biomass of 415 g · m$^{-2}$. The total aerial production during this cycle of growth initiated by the fire should be at least equivalent to the value of the maximum total biomass, since all of it was produced during the uninterrupted period of growth. We do not agree with the method of San José and Medina (1977) when they subtract from this total the biomass that was present at the time the rains started of 120 g · m$^{-2}$, which makes them estimate production at 650 g · m$^{-2}$, as if those 120 g · m$^{-2}$ corresponded to the previous growth cycle figured on the basis of a climatic cycle rather than of the biological cycle that was initiated by fire.

On the other hand, according to San José and Medina (1977), the unburned quadrat produced only 429 g · m$^{-2}$, that is, a value noticeably below that of the burned plot. On the basis of this difference, the authors maintain that burning stimulates a notable increase in productivity. The calculations and the theoretical concepts underlying this assumption appear dubious. The figure of 429 g · m$^{-2}$ for the production of the unburned savanna is obtained by subtracting from the peak of maximum total aerial biomass measured at the end of the growing season (1101 g · m$^{-2}$) the minimum value of the total aerial biomass calculated halfway into that season (672 g · m$^{-2}$ toward the end of July). When we apply our criteria to the data of San José and Medina, particularly curve 2A in figure 23, the total production can be calculated in two different ways which, however, give coincidental results. In the first place, taking into account all the increment of biomass that really occurred during the annual cycle considered, the total is not just the 429 g · m$^{-2}$ of the final growing period, but also the approximately 300 g · m$^{-2}$ accumulated between March and the end of May. This would add up to 730 g · m$^{-2}$, almost exactly the same figure as the estimate for the burned parcel by San José and Medina. Another way of estimating the annual production would be to add up the total of the decomposed annual biomass, which should be the same as the total biomass if there is no fire. The biomass decomposed in the interval between February and March (about 300 g · m$^{-2}$) plus the one decomposed between May and July (some 430 g · m$^{-2}$) again yields the figure of 730 g · m$^{-2}$ for the total biomass decomposed during the annual cycle.

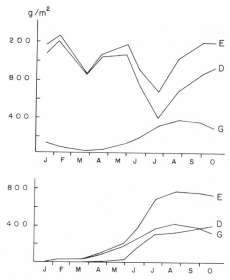

**Figure 23**  Annual variation of the green biomass *(G);* dry biomass *(D),* and aerial biomass *(E)* in a savanna of *Trachypogon plumosus* of the Biological Station of the llanos. *Upper graph:* parcel protected from fire; *lower graph:* parcel burned at the end of December. (Adapted from San José and Medina, 1977.)

In conclusion, our interpretation of the data of San José and Medina suggests that the only clear difference in productive processes between the burned and the protected savanna is that the former grows more green biomass. In other words, although the unburned savanna produces as much as the burned one, it shows greater mortality, and therefore at any time accumulates less green biomass. This greater mortality could be explained by assuming that the quantity of dead biomass accumulated in the protected savanna impedes the penetration of light and quickly turns the first leaves closest to the ground into inefficient photosynthetic tissues, which therefore dry out faster than in the burned savannas where more light comes through the canopy. This conclusion is quite different from the one that San José and Medina came to.

Their data provide the basis for the first credible estimate of the herbaceous aerial production of the savanna of *Trachypogon* in the region of Calabozo, calculated as a minimum value of $730 \text{ g} \cdot \text{m}^{-2}$ a year. On the other hand, the radical biomass in both parcels, to a depth of 30 cm, was $270 \text{ g} \cdot \text{m}^{-2}$ in the protected one and $170 \text{ g} \cdot \text{m}^{-2}$ in the burned one. On the assumption that fire did not modify the existing biomass, and that it was the same in both samples, it must be concluded that the underground production in the soil level tested is far greater in the protected parcel. Finally, the maximum growth rate in each savanna was almost the same: 11.0 g · m$^{-2}$ · day$^{-1}$ in the protected one and 10.5 g · m$^{-2}$ · day$^{-1}$ in the burned one.

Working at the Biological Station, Medina, Mendoza, and Montes (1977) measured the total aerial biomass in two savanna parcels of *Trachypogon plumosus-Axonopus canescens*, one burned in April 1973, the other cut with a mower. The experiment was repeated in 1974 in a savanna of *Trachypogon vestitus*, this time fertilized with potassium nitrate and potassium monophosphate. The rainfall in 1973 and 1974 was 1104.4 mm  and 1022.8 mm  respectively; that is, 15 to 20% below normal.

The total aerial biomass measured in November 1973 was 243.7 plus or minus 55.5 g · m$^{-2}$ in the mowed parcel and 326.1 ± 41.2 g · m$^{-2}$ in the burned parcel. The difference between them was significant at the 5% level. In this savanna the two dominant species accounted for 99% of the total biomass before the experiment; at the time of the final harvest, the proportion of *Trachypogon plumosus* in the aerial biomass of the burned quadrat was more than 96% of the total, while in the mowed parcel it was 86%. These data show an aerial production (not counting the period from November to the end of the cycle) of only 45% of what was established for the same community and with the same burning treatment in 1969 (326 versus 730 g · m$^{-2}$). This may be due to the great difference in rainfall in those years: rainfall in 1969 was 60% greater than in 1973. Another possible cause of the reduced production is the time when the parcels were burned: April, the date of burning in 1973, is much less favorable than the end of December, when it was done in 1969. The data also show that fire induces a superior growth than mowing, while the equilibrium among the two major dominant species is modified more by the latter treatment.

In 1974 the value of the maximal aerial biomass after the February fire was 280 g · m$^{-2}$, reached in November. This figure is still lower than that of the previous year, in spite of the earlier date of the burning; it was, however, a drier year. The fertilizer treatments increased the aerial production in a significant way, to 369 g · m$^{-2}$ (+N) and 331 g · m$^{-2}$ (+P). However, these values still are modest compared to 1969. In repeating the fertilization experiments with NPK. in the same savannas, San José and García Miragaya (1981) obtained an increase in herbaceous aerial production from 220 g $\cdot$ m$^{-2}$ for the control, to a maximum of 368 g $\cdot$ m$^{-2}$ for the land treated with the three nutrients. In the chapter about the economy of nutrients we will again discuss the results of fertilization treatments that modify the nutritional balance.

New data about seasonal changes in the subterranean biomass of the same *Trachypogon* savannas have been presented by San José et al. (1982). In the soil layer between the surface and 30 cm, where according to these authors 80% of the total below-ground dry weight is found, they observed seasonal oscillations of between 70 and 208 g $\cdot$ m$^{-2}$ in the unburned parcel, and for the parcel burned in March, between 125 and 167 g $\cdot$ m$^{-2}$. Basing their calculations on the seasonal distribution of live and dead roots, they conclude that the underground production of these parcels is 125 and 167 g $\cdot$ m$^{-2}$ respectively. Burning of the savanna thus leads to a significant increase in the growth of hypogeous material.

In addition to the *Trachypogon* savannas of Calabozo, San José and Medina also studied other savannas of the region. A community of flooded soils, dominated by species of *Axonopus*, reached a maximum value of 653 g $\cdot$ m$^{-2}$ of aerial biomass and 552 g $\cdot$ m$^{-2}$ of underground biomass. On the other hand, savannas on superficial soils over lateritic hardpans reached only 198 g $\cdot$ m$^{-2}$ and 229 g $\cdot$ m$^{-2}$ for the respective maximums of epigeous and hypogeous biomass.

Escobar and González Jiménez (1975, 1979) and González Jiménez and Escobar (1977) found in the ungrazed savannas of Hato El Frío (see fig. 2), a maximal aerial herbaceous biomass of 676 g $\cdot$ m$^{-2}$ in a community dominated by *Paspalum chaffanjonii*, *Axonopus purpusii*, and *Sporobolus indicus*, growing on the riverbank; of 745 g $\cdot$ m$^{-2}$ in communities dominated by *Panicum laxum*, *Paspalum chaffanjonii*, and *Leersia hexandra* in the lowlands; and of 857 g $\cdot$ m$^{-2}$ in a community dominated by *Hymenachne amplexicaulis* and *Leersia hexandra* in the marsh.

Bulla et al. (1977) studied the variation of the aerial biomass in savannas where drainage was regulated by dikes (modules) that allow the prolongation of the edaphic humid period. The work was done in Mantecal, in the state of Apure, very near the previous locality. In a topographic sequence that went from a lowland to a marsh, with maximal waterlogging conditons from 30 to 130 cm at the extremes of the gradient, the authors measured a maximal living biomass of 350 g $\cdot$ m$^{-2}$ and a maximal total aerial biomass of

780 g · m$^{-2}$ in the least muddy extreme, in the month of September; while at the other extreme of the saturation gradient, values of 700 g · m$^{-2}$ of green biomass and of 1380 g · m$^{-2}$ of total aerial biomass (in December–January) were obtained. In all cases two species of grasses, *Leersia hexandra* and *Hymenachne amplexicaulis*, together accounted for more than 95% of the total aerial biomass of this savanna.

Entrena and collaborators (1977), working in the same place and sequence, obtained a range of values for green biomass of between 370 and 600 g · m$^{-2}$ and for total aerial biomass, between 1000 and 1300 g · m$^{-2}$. Bulla et al. (1980a and b) present more complete results on the annual cycle of the aerial and underground biomass, production, and decomposition, in riverbank and marsh savannas. In the riverbank an annual peak of green biomass is reached in October (635 g · m$^{-2}$), for an annual production of 890 g · m$^{-2}$, and a decomposition value of 682 g · m$^{-2}$. Since in that year production was higher than decomposition, there was a net accumulation of standing dead vegetable biomass or of litter. In the same savanna the underground biomass oscillated between a minimum of 475 g · m$^{-2}$ in September and a maximum of 714 g · m$^{-2}$ in March. All these results apply to nonburned, protected parcels. The marsh savanna, which was not burned but stayed wet during the dry season by means of drainage control, had a maximum green biomass of 1719 g · m$^{-2}$, with an aerial production estimated through the sum of increments at 1097 g · m$^{-2}$ · year$^{-1}$.

These data indicate that the aerial biomass of riverbank and lowland savannas reaches similar values, which are in turn comparable to those obtained in the savannas of *Trachypogon* in very wet years but are higher than the average for these savannas. At the end of the annual cycle the swamp accumulates a greater green biomass, reaching maximal values of 1700 g · m$^{-2}$ when its herbaceous communities are free from water deficit throughout the year by drainage controls.

Escobar and Medina (1977) and Escobar (1977) present data on aerial and underground biomass in another floodable area, the *Paspalum fasciculatum* savanna of the marshes in the southwest of the state of Guarico. These sites become completely flooded during the rainy season as a result of the overflow of creeks and rivers. Escobar and Medina followed the evolution of the total biomass in three parcels, one unburned, one burned in April after the start of the rainy season, and the third one burned in March. After 90 days the respective biomass values were: 1300 g · m⁻², 900 g · m⁻², and 400 g · m⁻². The radical biomass found between the surface and 140 cm reached 1200 g · m⁻². The rate of decomposition was also measured, reaching in 90 days 60% for leaves, and 30% for stems. The annual aerial production, estimated through the sum of increments, was 1040 g · m⁻² · year⁻¹ for the burned parcel and 2540 g · m⁻² · year⁻¹ for the protected land. This community reaches a total aerial biomass and annual production superior to that of the other savannas subjected to flooding because the dominant species, *Paspalum fasciculatum*, is one of the largest grasses of the llanos.

Sarmiento and Vera (1978), working in the llanos of Barinas at a site located 16 km south of the city of Barinas and 7 km south of the Hato Caroni (see fig. 2), examined the annual production cycle of the total aerial biomass in three savannas that follow each other in a topographic sequence from an axis of embankment to the bottom of a broad and almost flat talweg. In previous studies (Silva and Sarmiento, 1976a and b) these three parcels were called savannas of Barinas, Garza, and Jaboncillo, names of the respective soil series of these parcels.

The parcel in Barinas was a grassland savanna with only one very isolated woody species *(Curatella americana)*; the dominant species in decreasing order were: *Axonopus purpusii, Leptocoryphium lanatum*, and *Trachypogon vestitus*. Several other species were also important; altogether 44 species were found in 100 m². The soil (Barinas series) is an oxic paleoustalf with deep hydromorphy (130 cm).

The savanna of Garza, a pure grassland dominated by *Paspalum plicatulum, Axonopus purpusii*, and *Trachypogon plumosus*, occupies an intermediate position in the sequence. The total number of species in 100 m² was 39. The soil (Garza series) is also an oxic paleoustalf, but with hydromorphy starting at 75 cm. In the lowest topographic position, some 150 cm below the embankment, lies the grassy savanna of Jaboncillo, dominated by *Sorghastrum parviflorum, Andropogon selloanus*, and *Leersia hexandra*. Here the total number of species found in 100 m² was 28. The soil is a typic tropaqualf, badly drained and periodically waterlogged during the rainy season (the water balance of these ecosystems will be discussed in the next chapter). The three parcels are burned each year during the first days of April as part of their normal management; they have also been subjected to very intensive grazing of approximately 7 ha · animal unit⁻¹.

The savanna of Barinas *(Axonopus purpusii-Leptocoryphium lanatum)* reached its maximum of total herbaceous aerial biomass toward the end of February (1974) with a value of $590 \pm 54$ g $\cdot$ m$^{-2}$. In the savanna of Garza *(Paspalum plicatulum-Axonopus purpusii)* a peak of maximal aerial biomass for the same period of $604 \pm 60$ g $\cdot$ m$^{-2}$ was obtained, and in that of Jaboncillo *(Sorghastrum parviflorum-Andropogon selloanus)* a value of $522 \pm 111$ g $\cdot$ m$^{-2}$ was obtained. (Error determined on the basis of three l m$^2$ samples.) The aerial growth during 1974–75 could only be followed in the communites of Garza and Jaboncillo, where at the end of the dry season respective maximal values of $552 \pm 27$ g $\cdot$ m$^{-2}$ and $705 \pm 33$ g $\cdot$ m$^{-2}$ were obtained.

The annual mean precipitation at the nearest meteorological station (Hato Caroni) was 1358 mm, but in the two years of the study the local values were only 1093 mm and 1017 mm. The difference in the maximal aerial biomass in the two years was not significant, and there was little difference between the three savannas in 1973; however, the difference between Garza and Jaboncillo in 1974 was significant.

The underground biomass of each savanna was measured on 1 November 1973 to 2 m of depth, with values of 963 g $\cdot$ m$^{-2}$, 816 g $\cdot$ m$^{-2}$, and 1100 g $\cdot$ m$^{-2}$ respectively. Later the variation between 0 and 20 cm was examined for Garza and Jaboncillo. The maximum values of total hypogeous biomass were obtained in February: 930 g $\cdot$ m$^{-2}$ and 1020 g $\cdot$ m$^{-2}$ respectively. The minimum values were obtained in the middle of the rainy season, in September: 384 g $\cdot$ m$^{-2}$ and 254 g $\cdot$ m$^{-2}$. Were the vertical distribution of underground biomass constant throughout the year, the extreme values over the total profile (0–200 cm) should be as follows: in the savanna of Garza, 1148 g $\cdot$ m$^{-2}$ and 474 g $\cdot$ m$^{-2}$, that is, an underground minimal production of 674 g $\cdot$ m$^{-2}$ $\cdot$ year$^{-1}$; in the savanna of Jaboncillo, 1357 g $\cdot$ m$^{-2}$ and 273 g $\cdot$ m$^{-2}$, corresponding to an underground production of 1084 g $\cdot$ m$^{-2}$ $\cdot$ year$^{-1}$. In summary, the aerial production estimated for the peak of maximal total biomass is in the order of 550 to 650 g $\cdot$ m$^{-2}$ $\cdot$ year$^{-1}$ at Garza and Barinas, and between 500 and 700 g $\cdot$ m$^{-2}$ $\cdot$ year$^{-1}$ in Jaboncillo. The underground production would be of the same order of magnitude as the aerial production at Garza, and greater (1084 g $\cdot$ m$^{-2}$ $\cdot$ year$^{-1}$) at Jaboncillo.

Another savanna under study was located at the place called Corozo-Palmitas, 20 km west of the city of Barinas (see fig. 2). The area has an annual mean precipitation of 1834 mm but also four months of drought. The years of our study, 1973 and 1974, turned out to be relatively dry with rainfall amounting to only 1330 mm and 1158 mm respectively. This savanna, which we called savanna of Boconoito, is an open savanna dominated by *Leptocoryphium lanatum* and *Elyonurus adustus*. It is a rich plant community, with 46 species in 100 m², several of them subshrubs. The researched parcel is burned every year at the beginning of February. The soil is a deep, well-drained, and relatively dry ustic haplustalf.

In 1973 the total maximal herbaceous aerial biomass was $534 \pm 14$ g $\cdot$ m⁻², and in 1974 it was $529 \pm 44$ g $\cdot$ m⁻², in both years measured toward the end of November since already in January, shortly before burning, the total biomass had diminished slightly. The hypogeous biomass found between the surface and 200 cm, measured at the beginning of November, was 1891 g $\cdot$ m⁻² (see fig. 9). The underground biomass on the level between 0 and 20 cm yielded a maximum of 830 g $\cdot$ m⁻² in February and a minimum of 415 g $\cdot$ m⁻² in September. The difference of 415 g $\cdot$ m⁻² allows the estimation of the hypogeous production at this edaphic level. If we convert these data into values for the entire profile to 200 cm, and assuming an annual constancy in the vertical distribution, an average hypogeous production of 560 g $\cdot$ m⁻² $\cdot$ year⁻¹ is obtained.

When the values of biomass and aerial and underground production are compared with those of the other three savannas analyzed in the same region, it can be appreciated that the differences are not great, with the values at Boconoito being somewhat lower (10 to 15%) than those at Barinas and Garza.

Research was also conducted in the savannas of the mesa de Barinas, in the occidental llanos. This site is a terrace of coarse pebbles of the early Pleistocene (Q4), elevated during the Andean uplift. The parcel was situated 5 km northwest of the city of Barinas, with a rainfall of 1498 mm $\cdot$ year⁻¹. The soil is ancient, highly fertilized, with a loamy/sandy/clay stratum of 30 to 60 cm covering a large deposit of quartz pebbles. This open savanna is dominated by *Trachypogon vestitus,* with *Axonopus canescens* and *Leptocoryphium lanatum* as subdominants. At the beginning of November 1977 the maximum total aerial biomass was $633 \pm 83$ g $\cdot$ m⁻² (on the basis of 5 samples of 1 m² each). At that time the hypogeous biomass to 60 cm was 1725 g $\cdot$ m⁻² (10 samples). Two months later, in January, with the dry season at its peak and shortly before the yearly burning, the total aerial biomass had diminished to $495 \pm 29$

**Table 4.** Maximum harvested herbaceous biomass and estimated minimal aerial production for several Venezuelan savannas. (The production estimates in this table do not necessarily agree with the estimates of the different authors.)

| Author, locality, and mean rainfall (mm) | Harvest year and month | Rainfall of harvest year (mm) | Dominant species in sample savannas | Treatment | Maximum aerial biomass (g/m²) | Estimated aerial production (g/m²/year) |
|---|---|---|---|---|---|---|
| Blydenstein (1962) Biological Station, Calabozo 1334 | 1961, Aug. | 1136 | *Trachypogon plumosus, Axonopus canescens* | No fire (2 years) | 680 | 241 |
| | 1961, Aug. | 1136 | *T. vestitus, A. canescens* | Fire, Dec. (1960) | 534 | 534 |
| | 1961, Aug. | 1136 | *T. plumosus, A. canescens* | Fire, Jan. (1961) | 229 | 229 |
| | 1961, Sept. | 1136 | *T. plumosus, A. canescens* | Fire, March (1961) | 230 | 230 |
| Blydenstein (1963) Biological Station, Calabozo, 1334 | 1962, Aug. | ca.1350 | *T. plumosus, A. canescens* | No fire (3 years) | 980 | 309 |
| | 1962, Aug. | ca.1350 | *T. plumosus, A. canescens* | Fire, Jan. (1961) | 520 | 252 |
| | 1962, Aug. | ca.1350 | *T. plumosus, A. canescens* | Fire, March (1961) | 590 | 378 |
| | 1962, Aug. | ca.1350 | *T. plumosus, A. canescens* | Fire, Nov. (1961) | 461 | 461 |
| | 1962, Sept. | ca.1350 | *T. plumosus, A. canescens* | Fire, March (1962) | 294 | 294 |
| | 1962, Sept. | ca.1350 | *T. vestitus, A. canescens* | Fire, Dec. (1960) | 850 | 308 |

| San José and Medina (1977) | 1969, Aug. | 1839 | T. montufari, T. plumosus | Fire, Dec. (1968) | 731 | 731 |
|---|---|---|---|---|---|---|
| Biological Station, Calabozo, 1334 | 1969, Aug. | 1839 | T. montufari, T. plumosus | No fire (5 years) | 1101 | 730 |
| Medina, Mendoza, and Montes (1977) | 1973, Nov. | 1104 | T. plumosus, A. canescens | Fire, April (1973) | 326 | 326 |
| Biological Station 1334 | 1974, Nov. | 1022 | T. vestitus, A. canescens | Fire, Feb. (1974). | 280 | 280 |
| | 1974, Nov. | 1022 | T. vestitus, A. canescens | Fire, Feb. (1974). Fertilized with N and K | 369 | 369 |
| | 1974, Nov. | 1022 | T. vestitus, A. canescens | Fire, Feb. (1974). Fertilized with P and K | 331 | 331 |
| Sarmiento and Vera (1978) | 1973, Feb. | 1093 | Leptocoryphium lanatum, T. vestitus | Fire, April (1973) | 590 | 590 |
| Hato Caroni, State of Barinas 1358 | 1973, Feb. | 1093 | Paspalum plicatulum, A. purpusii, T. plumosus | Fire, April (1973) | 604 | 604 |
| | 1973, Feb. | 1093 | Sorgastrum parviflorum, Andropogon selloanus, Leersia hexandra | Fire, April (1973) | 522 | 522 |
| | 1974, Feb. | 1017 | P. plicatulum, A. purpusii, T. plumosus | Fire, April (1974) | 552 | 552 |

**Table 4.** (*continued*).

| Author, locality, and mean rainfall (mm) | Harvest year and month | Rainfall of harvest year (mm) | Dominant species in sample savannas | Treatment | Maximum aerial biomass (g/m²) | Estimated aerial production (g/m²/year) |
|---|---|---|---|---|---|---|
| | 1974, Feb. | 1017 | *S. parviflorum, A. selloanus, L. hexandra* | Fire, April (1974) | 705 | 705 |
| Sarmiento and Vera (1978) Corozo-Palmitas n.a. | 1973, Nov. | 1330 | *L. lanatum, Elyonurus adustus* | Fire, Feb. (1973) | 534 | 534 |
| Sarmiento and Vera (1978) State of Barinas 1834 | 1974, Nov. | 1158 | *L. lanatum, E. adustus* | Fire, Feb. (1974) | 529 | 529 |
| Escobar and González Jiménez (1975) Hato El Frío, State of Apure | — | — | *Paspalum chaffanjonii, Axonopus purpusii, Sporobulus indicus* | Fire | 676 | 676 |
| 1600 | — | — | *Panicum laxum, Leersia hexandra, Paspalum chaffanjonii* | Fire | 745 | 745 |
| | — | — | *Hymenachne amplexicaule, Leersia hexandra* | Fire | 857 | 857 |

| Source | Species | Treatment | | | |
|---|---|---|---|---|---|
| Escobar (1977) State of Guarico 1600 | *Paspalum fasciculatum* *P. fasciculatum* | Fire No fire | — — | — — | 1040 2540 |
| Bulla et al. (1977) Mantecal, State of Apure 1674 | *P. chaffanjonii,* *A. purpusii,* *Sporobolus indicus* | No fire, water control, standing water, 30 cm. | — | 780 | 780 |
| | *P. chaffanjonii,* *A. purpusii, S. indicus* | No fire, water control, standing water, 30 cm. | — | — | 890 |
| Bulla et al. (1980a) Mantecal, State of Apure 1674 | *Leersia hexandra,* *Hymenachne* *amplexicaule* | No fire, water control, standing water, 140 cm. | — | 1380 | 1380 |
| Bulla et al. (1980b) Mantecal, State of Apure 1674 | *L. hexandra,* *H. amplexicaule* | No fire, water control, standing water, 140 cm. | — | — | 1097 |

$g \cdot m^{-2}$ and the underground biomass in the first 10 cm had diminished by 20%.

The most notable difference between the five savannas of the llanos of Barinas and the savannas of *Trachypogon* of the region of Calabozo is the greater floristic diversity of the former. Not only are there more species per unit of surface, but also there are no absolute dominants like the species of *Trachypogon* in the savannas of the Biological Station, where they constitute more than 80% (sometimes up to 96%) of the total biomass. In the llanos of Barinas there are always two or three dominant species with several others also reaching relatively high biomass values (see fig. 8). Since these dominant species are unsynchronized in their annual growth and reproductive life cycle, they have a more regular increment in biomass during the annual cycle, so that the assimilatory process is prolonged almost to the end of the dry season. In contrast, in the savannas of Calabozo the peak of maximum total biomass is already reached in August, much before the end of the favorable season. The savanna of *Trachypogon vestitus* of the mesa of Barinas is the one that reaches its maximum total biomass earliest (beginning of November); it is therefore intermediate in this character between the other savannas of Barinas and the savannas of *Trachypogon* of the central llanos.

In floristic richness the savannas of the embankments and low-lands of Apure are intermediate between those of Calabozo and Barinas; however, some low areas, and marshes are similar to the savannas of Calabozo in that they are essentially integrated by two species, *Leersia hexandra* and *Hymenanche amplexicaulis*.

Table 4 summarizes all the information available about biomass and aerial production in the savannas of Venezuela. In the savannas of *Trachypogon* of Calabozo, the measurements performed by different researchers have resulted in figures for the maximum value of the annual biomass that range from 229 $g \cdot m^{-2}$ to 730 $g \cdot m^{-2}$; so that if all measurements are taken into account, including burned and not burned savannas but excluding mowed or fertilized savannas, a mean value of 379 $g \cdot m^{-2} \cdot year^{-1}$ for the 14 measurements is obtained.

But is is only in Calabozo that comparisons between treatments and years were made, and production was related to variations in the dominant species. Although this information is still inconclusive, the data tend to show a correlation between the values of the aerial biomass and the amount of rainfall. The timing of the burning is also an influencing factor, and it appears that early burning (November and December) increased production in comparison with later burnings (January to April). Whether or not burning increases total aerial production is not certain, but it does seem to increase the green biomass accumulated at the peak of total biomass. Fertilization increased production, but the moderate quantitites of fertilizer yielded only moderate increases. Finally, there seems to be no difference in production among communities with different dominant species (*Trachypogon montufari*, *T. plumosus*, or *T. vestitus*).

Another interesting phenomenon is the progressive accumulation of epigeous matter in the savannas protected from fire during several consecutive years. The values obtained by Blydenstein and by San José and Medina are shown in table 5. The quantity of accumulated epigeous biomass keeps increasing yearly until it stabilizes at values close to 1200 g $\cdot$ m$^{-2}$. That is, the biomass produced in a given year is not totally decomposed during that developmental cycle nor during the following one, but a certain quantity of dead standing biomass accumulates until a balance between production and decomposition is reached after four or five years.

In the savannas of the llanos of Barinas, production values for all communites and years range from 522 g $\cdot$ m$^{-2}$ $\cdot$ year$^{-1}$ to 705 g $\cdot$ m$^{-2}$ $\cdot$ year$^{-1}$, with a mean for the eight measurements of 584 g $\cdot$ m$^{-2}$ $\cdot$ year$^{-1}$. The annual production of these savannas is thus significantly greater than that of the savannas of *Trachypogon* of the central llanos, even though in the years of measurement rainfall

**Table 5.** Maximum epigeous biomass in a protected savanna of *Trachypogon* of the Biological Station of the Llanos, during five years. Data from Blydenstein (1962, 1963); San José and Medina (1977); and Medina et al. (1977).

| Time since last fire | Maximum aerial biomass (g/m²) |
|---|---|
| 1 year | 230–730 |
| 2 years | 520–850 |
| 3 years | 980 |
| 4 years | 1200 |
| 5 years | 1200 |

**Table 6.** Maximum and minimum underground biomass and estimate of the underground production in various Venezuelan savannas.

| Author | Depth (cm) | Maximum biomass (g/m²) | Minimum biomass (g/m²) | Estimated production (g/m²/year) | Year | Dominant species in sample savannas |
|---|---|---|---|---|---|---|
| | | | *Biological Station, Calabozo* | | | |
| Blydenstein (1962) | 0–10 | 162 | — | — | 1961 | *T. plumosus, A. canescens* |
| | 0–10 | 243 | — | — | 1961 | *T. plumosus, A. canescens* |
| | 0–10 | 292 | — | — | 1961 | *T. vestitus, A. canescens* |
| San José and Medina (1977) | 0–30 | 270 (protected) | — | — | 1969 | *T. montufari, T. plumosus, A. canescens* |
| | 0–30 | 170 (burned) | — | — | 1969 | *T. montufari, T. plumosus, A. canescens* |
| San José et al. (1982) | 0–30 | 208 | 70 | 194.8 | — | *T. plumosus, A. canescens* |
| | | | *Southeast of Guarico* | | | |
| Escobar and Medina (1977) | 0–140 | 1200 | — | — | 1976 | *Paspalum fasciculatum* |
| | | | *Hato Caroni* | | | |
| Sarmiento and Vera (1978) | 0–20 | 930 | 384 | 546 | 1974/75 | *Paspalum plicatulum, Axonopus purpusii* |
| | 0–200 | 1148 | 474 | 674 | 1974/75 | *P. plicatulum, A. purpusii* |
| | 0–20 | 1020 | 254 | 766 | 1974/75 | *Sorghastrum parviflorum, Andropogon selloanus* |
| | | | *Corozo-Palmitas* | | | |
| | 0–200 | 1357 | 273 | 1084 | 1974/75 | *Leersia hexandra* |
| | 0–20 | 1400 | 415 | 985 | 1973/74 | *Leptocoryphium lanatum, Elyonurus adustus* |
| | 0–200 | 1891 | 561 | 1330 | 1973/74 | *L. lanatum, E. adustus* |

was below normal in the area, and about the same as in Calabozo. The effect of fire, of timing of the burn, and of yearly rainfall variations cannot be ascertained from the available data. On the other hand, there are no significant variations between communities, excepting perhaps the hyperseasonal savanna of *Sorghastrum parviflorum* (Jaboncillo). It must be pointed out also that in Calabozo the *Andropogon* savanna in floodable soils produced a greater biomass than the *Trachypogon* savanna in well-drained soils, and that the floodable savanna of *Paspalum fasciculatum* has apparently the most productive herbaceous community studied in Venezuela.

Finally, table 6 gathers up the available information about biomass and hypogeous production. In the savannas of *Trachypogon* in the llanos of Calabozo, the hypogeous biomass in the horizon to 30 cm is of the same order of magnitude as the maximum epigeous biomass in a year (200 to 300 g $\cdot$ m$^{-2}$); while the savannas of the llanos of Barinas have a much greater maximum hypogeous biomass: 900 to 1400 g $\cdot$ m$^{-2}$ in the first 20 cm (1100 to 1900 g $\cdot$ m$^{-2}$ if the whole profile is taken into account). The floodable savannas of the south of the state of Guarico also have a large underground biomass. The hypogeous production estimated from the difference between the maximum and the minimum of the underground biomass in the savannas of the western llanos reaches values that are similar or slightly superior to those for the epigeous production at the peak of aerial growth.

# Methodological and conceptual problems in the study of productive processes

An exhaustive critical survey of the different methodologies used to determine primary production in herbaceous communities has already been presented in recent publications (Milner et al., 1968; Whittaker and Marks, 1975; Singh et al., 1975). Here it is only necessary to point out briefly the main deficiencies of the methods used in order to speculate on the possible differences between the production estimates for grasslands and savannas published in the literature and the values that would be obtained if each of the unitary processes that are part of the complex process of primary production could be measured separately.

Leaving aside methods that are just being tried, of direct measurement of gas exchange in intact communities, the most common method used until now to determine aerial or underground production during a vegetative cycle is the so-called direct method of successive harvests, which is especially appropriate in herbaceous communities where most of the biomass is renewed annually. The simplest way is to harvest the biomass at periodic intervals and to consider that production is equivalent to the largest harvest, the so-called annual peak of the biomass, provided one starts from nothing as is the case in a burned savanna. In the case of a protected savanna that already had some accumulated biomass at the beginning of the yearly cycle, it is necessary to calculate the difference between the maximum and the minimum.

This way of estimating production has major deficiencies. First of all, there may be various species that peak at different times, resulting in an important underestimation of annual productivity. In that case it would be more precise to add the annual peaks of each species, or at least of the major ones; otherwise, the value calculated for the annual production will not include a part of the vegetative growth and a great part of the reproductive biomass of those species that attain their zenith at another time of the year. Malone (1968) calculated that in a seral herbaceous community of the temperate zone, where the species reach their maximum at different times of the year, taking the annual peak as total production results in an underestimate of 100%. However, it is very time-consuming to make the determination species by species in floristically mixed communities; also, sampling of the less abundant ones is always insufficient unless one takes a very large number of samples.

For example, in the savannas of *Leptocoryphium lanatum-Elyonurus adustus* (Boconoito) of the foothills of Barinas (Sarmiento and Vera, 1978), the maximum aerial biomass is reached toward the end of November (534 g · m$^{-2}$); but one of its dominant species, *Elyonurus adustus* completes all of its life cycle between February and April, so that by November there are not even vestiges of the inflorescences, spikelets, or seeds of this species. The biomass of these structures, measured at the time of maximum development, reached 20 g · m$^{-2}$. The same is true for other dominant or subdominant species of this community — *Leptocoryphium lanatum* blooms in April–May, *Sporobulus cubensis* in March–April, *Bulbostylis junciformis* in July–August, *Trachypogon plumosus* in August–September. A conservative estimate would add over 100 g to the maximum for this ecosystem as calculated in November.

Second, the method of the successive harvests does not take into account the opposite processes of net assimilation and decomposition that take place simultaneously during much of the year. That is, in considering the peak of maximum biomass as the annual production, the assumption is made that production takes place until the peak of biomass is reached; then assimilation stops, previously produced matter dies, and once that process is ended the decomposition of the dead biomass begins. Consequently there must be no assimilation after the peak of green matter is reached, nor can there be decomposition before the peak of dry matter is reached. These assumptions are, however, incorrect. In tropical savannas the individual plants usually produce new structures or develop existing ones simultaneously with the death of other tissues and organs and the decomposition of the dead matter, whether on the plant as is generally the case with savanna grasses, or as part of the litter on the soil surface. Therefore, in order to estimate production correctly, it is necessary to determine the rate of decomposition and assimilation throughout the year. Likewise, in order to estimate total production, translocation of organic matter between shoot and root must be taken into account. Different techniques exist to estimate decomposition and mortality (see for example Wiegert and Evans, 1964; Lomnicki et al., 1968), based either on the determination of the decomposition of vegetable dry matter placed on the ground in open mesh plastic, or on the periodic elimination of the dry matter in quadrats, thereby measuring mortality at successive time intervals. If the dead biomass is also sorted out species by species, at sufficiently short intervals (2 to 4 days), the mortality rate can be calculated with no measurable error caused by decomposition of the dry matter; this is what Hunt (1970) did in an artificial two-species prairie, but the method is not easily applicable to natural herbaceous communities rich in species. In any case, there is very little data on these parameters and their variation in the annual cycle in tropical grasslands and savannas.

## A model for the analysis of productive processes

The productive process consists of several opposing processes that take place simultaneously. The author and his colleagues have developed a simple model adapted to the available information, which allows the monitoring of these phenomena in the herba-

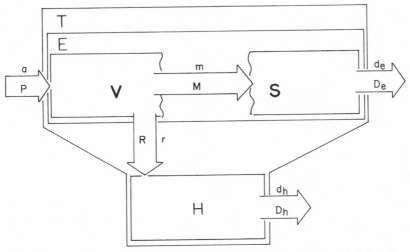

**Figure 24** Schematic representation of the productive model for the savanna. See table 7 for the principal characteristics of the model. $T$: total biomass; $E$: above-ground biomass; $H$: underground biomass; $V$: live above-ground biomass; $S$: dry above-ground biomass; $P$: net primary production; $a$: assimilation rate; $M$: mortality; $m$: mortality rate; $R$: reallocation; $r$: reallocation rate; $De$: above-ground decomposition; $de$: rate of above-ground decomposition; $Dh$: below-ground decomposition; $dh$: rate of below-ground decomposition.

ceous cover of a savanna throughout an annual cycle. The model intends to clarify the various processes — the input and output of the system — and make explicit the assumptions and premises on which the model has been established.

The global productive system is formed by three compartments or subsystems (fig. 24). $V$ corresponds to the live epigeous biomass, which in a grassland corresponds closely with the green or photo-synthetic biomass; $S$ is the dry aerial standing biomass; $H$ corresponds to the total underground or hypogeous biomass. In this case no distinction is made between living and dead biomass because of the substantial operational difficulties that are involved in such a separation. The reproductive biomass has not been considered separately in this model, and it will be placed in either $V$ or $S$. The set $V + S$ forms the epigeous subsystem ($E$) within the total system, $T$.

Five different functional processes or rates of organic matter conversion connect the subsystems as follows:

The production $P$ is the total quantity of biomass incorporated into the system $T$, in a given time interval, through the assimilatory system $V$. We here consider the net production of the community in the sense of Odum (1971), that is the primary net production less maintenance. We will call $a$ the mean assimilation rate $(g \cdot m^{-2} \cdot day^{-1})$ during a given time interval. $P$ is the only entry point into the system.

The mortality $M$ is the quantity transferred from $V$ to $S$ in a given time interval at a mean rate of $m$ in $g \cdot m^{-2} \cdot day^{-1}$. $M$ is the only entry into compartment $S$.

The epigeous decomposition $De$ is the total quantity of standing dead organic matter, which disappears during a certain time interval at a rate of $de$ in $g \cdot m^{-2} \cdot day^{-1}$. In this model a separate compartment for the litter was not considered, because it is assumed that the dry biomass, formed mainly by grasses and sedges, decomposes as standing biomass until fire destroys it totally.

The reallocation $R$ is the net quantity of organic matter that was interchanged between $V$ and $H$ at a rate of $r$ in $g \cdot m^{-2} \cdot day^{-1}$. In principle it can have positive values when the net reallocation is toward the hypogeous organs, or negative values if the net reallocation is toward the aerial organs, as is the case during the early stages of development in tropical savannas. Note also that in referring to "hypogeous production" one is committing a verbal inaccuracy, since in reality what is meant is that part of production (necessarily epigeous) is transported or reallocated toward $H$.

The hypogeous decomposition $Dh$ is the only output of the subsystem $H$, at a mean rate of $dh$ in $g \cdot m^{-2} \cdot day^{-1}$ during a given time interval.

The model can even be simplified, especially if there is no information about the annual changes in the hypogeous biomass. In that case compartment $H$ can be eliminated and processes $R$ and $Dh$ disappear, while total production $P$ is changed into epigeous production $Pe$.

Of all these variables and rates of transfer that take place in the productive process of a herbaceous community, we can measure directly through the method of successive harvests the variations in $V$, $S$ and $H$ (these last with a large sampling error) and the rates of decomposition, although techniques for it are somewhat deficient. The assimilation rate $a$ can also be measured experimentally, but at too short an interval compared with the annual cycle; hence it is usually calculated from the other variables. It is also possible to measure independently the death rate $m$.

Table 7 summarizes the components and the principal properties of the system and the relations between the various processes.

# The productive processes of the tropical savannas

## Seasonal rhythms

Production in the neotropical savannas has been defined as equivalent to the difference between the maximum and minimum values of the annual biomass, treating the above- and below-ground components separately. This difference will give a correct estimate of the annual production only when there has been no decomposition of the biomass produced during that cycle. Only under the following conditions could there be parity between production and biomass increment:

$$Pe = E = V + S$$
$$Ph = H$$
$$Pt = E + H$$

Clearly, however, it is incorrect to assume that there is no decomposition. Indeed, observation of a seasonal savanna burned during the last weeks of the dry season reveals that very soon after the fire a phase of active growth starts for the majority of the perennial species, both herbaceous and woody, which translates in the blooming and fruiting of the precocious species and in the rapid vegetative growth of all species.

Four to eight weeks after burning, or a little later if burning took place early, all the precociously blooming species have finished their reproductive cycle, with only a few inflorescences or fruits remaining. At the same time they have developed a great part of

**Table 7.** Some properties of the model of the productive processes of savanna ecosystems (fig. 24).

1. Increment of $V$ in time $t$, starting from $V_0$:
   $$I_V = V_t - V_0 = a \cdot t - r \cdot t - m \cdot t$$

2. Increment of $S$ in time $t$, starting from $S_0$:
   $$I_S = S_t - S_0 = m \cdot t - d_e \cdot t$$

3. Increment of $E$ in time $t$, starting from $E_0$:
   $$I_E = E_t - E_0 = a \cdot t - d_e \cdot t - r \cdot t$$

4. Increment of $T$ in time $t$, starting from $T_0$:
   $$I_T = T_t - T_0 = a \cdot t - d_e \cdot t - d_h \cdot t$$

5. Increment of $H$ in time $t$, starting from $H_0$:
   $$I_H = H_t - H_0 = r \cdot t - d_h \cdot t$$

6. Mean rate of assimilation in time $t$:
   $$a = I_T/t + d_e + d_h = I_E/t + d_e + r$$

7. Total production in time $t$:
   $$P = a \cdot t = I_T + d_e \cdot t + d_h \cdot t = I_E + d_e \cdot t + r \cdot t$$

8. Mean mortality rate in time $t$:
   $$m = I_S/t + d_e = -I_V/t - r$$

9. Epigeous mortality in time $t$:
   $$M = I_S + d_e \cdot t = I_S + D_e$$

10. Mean rate of decomposition of the aerial biomass in time $t$:
    $$d_e = -I_E/t = -I_S/t + m$$

11. Total decomposition of the aerial biomass in time $t$:
    $$D_e = d_e \cdot t = -I_E = -I_S + M$$

12. Mean rate of decomposition of the hypogeous biomass in time $t$:
    $$d_h = -I_H/t + r$$

13. Total decomposition of the hypogeous biomass in time $t$:
    $$D_h = d_h \cdot t = -I_H + r \cdot t$$

14. Mean rate of translocation during time $t$:
    $$r = I_H/t + d_h$$

15. Total amount reallocated from the aerial to the underground organs in time $t$:
    $$R = I_H + d_h \cdot t = r \cdot t = I_H + D_H$$

16. Epigeous production during time $t$:
    $$P_E = P_T - P_H = I_V + I_S + D_E = I_V + M = a \cdot t - r \cdot t$$

17. Annual hypogeous production in a savanna in equilibrium:
    $$P_H = R = r \cdot t = I_H + D_H = I_H + d_h \cdot t$$

18. Annual epigeous production in a savanna in equilibrium:
    $$P_{E \cdot year} = M_{year} = D_{E\,year}$$

19. Total annual production in a savanna in equilibrium:
    $$P_{year} = M_{year} + R_{year}$$

their assimilatory apparatus, so that the epigeous vegetative growth during the rest of the year will be less important. These species end their reproductive activities two or three months after the start of the active cycle, but they remain completely green except for the inflorescences.

Meanwhile, the species with a late phenorhythm have been developing much more slowly. Thus in those savannas where species with both types of phenological behavior coexist, this type of behavior leads temporarily to a complete aspect dominance (biovolume) by the precocious species. For example, the savannas of the mesa of Barinas are normally burned in March. From then until the beginning of May they seem to be dominated by precocious species such as *Elyonurus adustus, Leptocoryphium lanatum, Sporobulus cubensis, Bulbostylis paradoxa,* and several Papilionoideae. Only after May do the late species begin to advance, and by July the balance appears equal between the precocious and late species (*Trachypogon spp., Andropogon spp.,* and *Axonopus spp.*). After July, when the reproductive development of the late species starts, they dominate the system physiognomically.

At that point the precocious species have all suffered a more or less severe mortality process and by the middle of the wet season have a high proportion of dead aerial biomass. Not only is the rate of epigeous mortality high, but also part or all of the aerial biomass of some precocious species has begun to decompose. By the time the late species reach their peak, the majority of the geophytes have disappeared from the surface. The late species also begin to dry out after fruiting, so that by the beginning of the dry season about half of the epigeous biomass is completely dried out (see figs. 7 and 8). During the four or five months of the dry season, until the next burning or until the first rains if there is no fire, the proportion of standing dry biomass increases constantly. When a new annual growth cycle begins, the green biomass of this savanna constitutes only about 10% of the total aerial biomass.

It is only logical to assume that during this whole period the dry biomass has been decomposing, even though the rate of decomposition per unit of dry weight must remain relatively low during the dry season. It will only increase steeply with the advent of the next rainy period, provided it has not been burned before then.

In summary, assimilation in the savannas is a continuous process during the whole annual cycle, although of course it goes on at different rates in different periods. This is true whether we consider the herbaceous stratum as a whole or each of the different

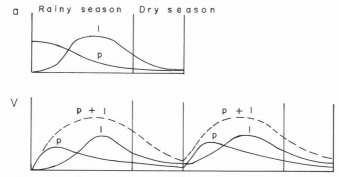

**Figure 25** Annual variation of some parameters in the productive process of a tropical savanna. Because the vertical axes are relative, no numerical values are given. The *top* graph represents the changes in the rates of assimilation *(a)* in perennial precocious *(p)* and late *(l)* species. The *bottom* graph indicates the changes in the green biomass and the sum of both. For greater clarity two annual cycles have been depicted.

phenological productive groups separately. Assimilation and growth rates are very high in the precocious species during the first two rainy months and relatively modest in the late species during that same period. After that the proportion is inverted until the reproductive cycle of the late species is completed. From that moment on assimilation declines in all the species, but it continues nevertheless until the end of the cycle. This is attested by the existence of an assimilatory biomass, as well as by the phenological behavior of the group of species that only go into semidormancy during the unfavorable season. Figure 25 shows qualitatively the annual variation of the green biomass and the assimilation rates for each species in these two groups.

There appears to be no mortality during the first two or three months of growth. But once the reproductive cycle of the precocious species is completed, mortality starts to accelerate and does not stop until the annual cycle is completed. Epigeous decomposition is certain to follow the cycle of mortality, as many precocious species become totally decomposed shortly after completing their reproductive phase. The reproductive structures of the perennial grasses of early flowering decompose shortly after seed dispersal, and by the middle of the wet season they have totally disappeared; in like manner the annuals decompose throughout the season, and nothing is left of them by the time the dry season arrives. The late blooming perennials, which decompose more slowly because they reach their vegetative and reproductive peak toward the end of the rainy season, just before the decomposition processes slow down during the dry season, are also affected by decomposition processes as is attested by the quick disappearance of the reproductive structures after seed dispersal.

Very little is known about the processes of reallocation and hypogeous production in tropical savannas. However, one can extrapolate from some results obtained in temperate grasslands (Troughton, 1957; Golley, 1965; Kelly, 1975). Apparently there is an inverse relation between aerial and underground development in prairies and other herbaceous communites in the temperate zones, where the annual growth cycle extends from spring to fall, while during the winter the species enter into a phase of semirest or dormancy (cryopause). In spring the first phase of aerial and radical development begins; the latter appears to stop temporarily during the summer when the species enter into their reproductive phase. Toward the end of the summer and during the fall the underground development starts again, and not only most new photosynthates, but also reserve products and the nutrients from senescing leaves and shoots are directed toward the hypogeous organs. The same phenomenon of massive translocation occurs in tropical savannas at the beginning of the dry season (see chap. 6).

The above observations of seasonal productive processes in tropical savannas lead to the deduction that equating the maximum green biomass, which occurs towards the middle of the wet season, with annual production implies a subestimation of the same for the following reasons:

1. Assimilation is continuous during the year, so that the accumulation of epigeous (and hypogeous) biomass continues between the time of maximum biomass and the end of the cycle.

2. The individual species maxima of green biomass are not synchronized with that of the community.

3. Aerial tissue mortality starts before the maximum of green biomass is reached, so that at that moment a certain amount of the standing biomass produced during that cycle is already dry.

4. Decomposition also starts before the peak of green biomass is reached, so that a proportion of the dry biomass of that cycle has already disappeared, particularly that of the precocious species.

It can similarly be established that the peak of maximal aerial biomass (dry plus green), which in different savannas is reached between the middle and the end of the dry season, does not represent a correct estimation of the epigeous production either, since by then a considerable amount of the aerial dry biomass has already decomposed. To obtain more satisfactory estimates one needs to have measurements of the rates of aerial and underground decomposition, calculated separately for the different organs such as shoots, leaves, inflorescences, roots, and rhizomes, since they grow and decompose at different speeds and rhythms during the year. Unfortunately the rates of decomposition for neotropical savannas are not known. To obtain a first approximation we can make use of two sources of information: the rate of decomposition of dead matter accumulated on unburned savannas; the rates of decomposition measured in other tropical savannas.

On the basis of the curves of the annual variation of the green and dry biomass formulated by San José and Medina (1977) for the *Trachypogon* savannas of the Biological Station, we have calculated the minimal rates of decomposition (on the assumption that there is no assimilation during these periods). At the end of the dry season (February and March) the mean rate of decomposition was 7.6 $g \cdot m^{-2} \cdot day^{-1}$ with a total rainfall for that period of 41.2 mm; while in June-July the corresponding mean rate of decomposition was 13.6 $g \cdot m^{-2} \cdot day^{-1}$ with a total rainfall of 689.3 mm. Since it may be assumed that in this savanna there is an equilibrium between production and decomposition after being protected from fire for five years, the total amount of biomass produced in a year must equal the amount decomposed. Consequently, and assuming that decomposition took place only during the two periods mentioned — the only ones where these calculations can be made as a result of actual decrease in biomass — the total epigeous decomposition would be 1060 $g \cdot m^{-2} \cdot year^{-1}$. Evidently decomposition must have been taking place throughout the year, although at a lower rate during the dry season. Consequently decomposition may easily reach 1500 $g \cdot m^{-2} \cdot year^{-1}$, or twice the peak of total maximal biomass (730 g). We should add that this particular year was an exceptionally wet one, and therefore the production would have been much greater than in less rainy years. For the same reason it could be assumed that the total value of decomposition has also been exceptionally high. However, that was not the case, since the total biomass of the savanna has not changed significantly between the beginning and the end of this cycle (1200 and 1170 g · m$^{-2}$).

From the calculation of the decomposition of the biomass values of aerial production are obtained that are up to 100% higher than those measured at the peak of the biomass, indicating that approximately half of the annual production decomposes during the same growth cycle in which it is produced. In the embankment savanna of Mantecal, Bulla et al. (1980a) measured the rates of decomposition during the rainy season. These oscillated between 3 and 7 mg · g$^{-1}$ · day$^{-1}$, for a total annual decomposition rate of the aerial biomass of 682 g · m$^{-2}$. If aerial production had been calculated through the increment of the biomass method, it would have been estimated at 520 g · m$^{-2}$. From the rates of decomposition, however, the value obtained was 890 g · m$^{-2}$, that is, 70% over the first figure.

It is worth mentioning that for a savanna of *Hyparrhenia rufa* in Costa Rica Daubenmire (1972) determined an annual decomposition equivalent to 50% of the maximum biomass. Cesar and Menaut (1974), working in two savannas in Lamto, Ivory Coast, determined monthly decomposition rates of 7.6% and 12%, which would increase the annual production between 30 and 55% above the biomass peak. But the decomposition values obtained from the Venezuelan savannas of *Trachypogon* are probably closer to the general values of New World savannas, so that the aerial production and the annual decomposition in the unburned savannas may be assumed to be between 50 and 100% higher than the annual peaks of biomass.

The greatest difficulty still lies in obtaining a more accurate estimate of the underground production than that derived from the difference between the maximum and the minimum of the hypogeous biomass. This is undoubtedly the weakest point of all in the analysis of the productive system of the tropical savanna. Not only do the determinations of the underground biomass suffer serious sampling and methodological problems, but there are very few estimates of the annual rates of mortality and decomposition. Even in other herbaceous systems the hypogeous data are insufficient and incapable of giving a picture of the dynamics of the underground biomass.

Weaver and Zink (1946) conducted a now classical study on the survival of the underground biomass of Gramineae of the American prairie, cultivated in the laboratory over three years. Analyzing the

behavior of ten species, these authors find great interspecific variations in mortality and in the rhythms of disappearance of the roots
in successive years. Their data show that the renewal of the roots of
these perennials is not an annual phenomenon, as some researchers had maintained. On the contrary, a good proportion of
the roots lives more than a year; moreover, at the end of two years
all the species had different proportions of live roots of that age.
That is, these prairie grasses renew their roots in cycles longer than
two years.

Dahlman and Kucera (1965) studied the underground production
of a prairie in Missouri and obtained a maximum value for the
hypogeous biomass of 1900 g · m$^{-2}$. They determined a hypogeous production of about 450–500 g · m$^{-2}$ · year$^{-1}$, equal to the
annual decomposition of this prairie which has reached equilibrium, which implies that approximately one fourth of this quantity
is renewed annually. That is, the renewal of the hypogeous biomass
occurs every four years, with a yearly renewal rate of 23% for the
rhizomes and 42% for the roots. But some roots live for periods
shorter than a year. Their mortality and corresponding decomposition thus do not figure in the calculations of the annual production, which would then be underestimated, and the recycling
period of the underground biomass would thus be less than four
years.

When we measured the hypogeous biomass of various communities in the savannas of the western llanos (Sarmiento and Vera,
1978), we obtained maximum values of between 1148 and 2000
g · m$^{-2}$ and annual differences of hypogeous biomass between 674
and 1330 g · m$^{-2}$. That is, even without considering the decomposition of the hypogeous biomass that takes place in the cycle of its
formation, at least 60 to 75% of the hypogeous biomass is renewed
every year, which means much higher recycling rates than in the
temperate regions. Taking into account the large proportion of
rhizomes within the hypogeous biomass, it would mean that almost
all of the roots live only a year. San José et al. (1982) arrive to a
similar conclusion, that is, that the annual growth is equal to the
amount of decomposed dead matter in the soil.

## Estimate of annual production

The discussion of the various productive processes in tropical
savannas would seem to warrant the hypothesis that the net annual

epigeous production in these ecosystems is at least 50% higher than the annual biomass peak. Likewise it may be assumed that the hypogeous production will be higher than the yearly increment in underground biomass by a yet undetermined amount.

The western llanos of Venezuela in their vegetation and environment probably represent the mean values for neotropical savannas. A summary of all our data would offer the following picture of net primary production in grams per square meter:

|  | Epigeous | Hypogeous | Total |
|---|---|---|---|
| Seasonal savannas | 800 – 1300 | 600 – 800 | 1400 – 2100 |
| Hyperseasonal savannas | 800 – 1400 | 900 – 1100 | 1700 – 2500 |

These values represent the range within normal ecological conditions. They tend to increase as the total annual precipitation increases. They also change, in not yet clearly determined ways, with fire protection, as well as with the timing of the fire.

As for the production of the woody stratum, it is much lower than that of the herbaceous stratum, even in closed savannas where the woody biomass is much greater than the herbaceous one. For the Venezuelan llanos Vera (1978) has elaborated data on the production of leaf litter in a woody savanna in the foothills of the llanos of Barinas. In this ecosystem six woody species — *Curatella americana, Byrsonima crassifolia, Byrsonima coccolobaefolia, Roupala complicata, Bowdichia virgilioides* and *Palicourea rigida* — constitute most of the biomass of this stratum, with a density of 1000 trees per hectare and a cover of the woody stratum of 30%. The annual production of leaf litter reached 1500 kg $\cdot$ ha$^{-1}$. This amount plus the annual growth of the aerial organs of the trees would yield the value of the aerial annual production of this stratum. As there is no direct measure of the biomass increment of the trees, it is necessary to use the relationship between leaf and wood, valid for all tropical formations, elaborated by Huttel and Bernhard-Reversat (1975). We have estimated that the formation of wood is approximately 50% of the annual leaf production, which would give a production value of 200 g $\cdot$ m$^{-2}$ for the woody stratum, or about 20 to 25% of the epigeous production of the herbaceous stratum. We have no way to estimate either the actual amount or the order of magnitude of the hypogeous production of the trees.

## Time needed to reach equilibrium

The various data for the biomass in the *Trachypogon* savannas of the Biological Station of the Llanos can also be used to estimate the time that this ecosystem needs to reach a steady state, since the Biological Station is the only privileged place that can protect savannas from burning and keep them under observation for more or less prolonged periods. As table 5 shows, at the end of four consecutive years without fires this savanna appears to reach a state of equilibrium around 1200 g · m⁻². From that point on, each year as much biomass is decomposed as is produced. But during the first four years after burning epigeous decomposition was less than production, so that the biomass kept accumulating in the form of dry, undecomposed, standing litter. In a state of equilibrium the epigeous biomass shows small oscillations during each annual cycle, because at the beginning of each growth cycle production is greater than decomposition. This is reversed later. Figure 26 depicts the oscillations of the green, dry, and total biomass in each cycle as well as throughout the whole period from the start of growth following a fire until the steady state is reached.

After the dominant species enter a secondary succession, they gradually develop their perennating organs, incrementing their biomass year by year up to the point where decomposition of dry and old tissue equals production. This probably takes three or four years.

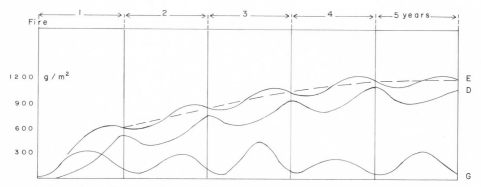

**Figure 26**   The variations in the green *(G)*, dry *(D)*, and above-ground *(E)* biomass of a tropical savanna from the time it is burned until it reaches equilibrium in five years. The *dashed* line indicates the tendency towards stabilization of the maximal above-ground biomass after four years of growth.

## Production rates and recycling times

The mean daily values of the daily production rates, computed on the basis of the lowest production figures within the previously established range, are:

Mean annual assimilation rate per unit of surface:

$$a = \frac{1400 \text{ g} \cdot \text{m}^{-2}}{365 \text{ days}} = 3.8 \text{ g} \cdot \text{m}^{-2} \cdot \text{day}^{-1}.$$

Mean annual aerial mortality rate per unit of surface = mean annual epigeous decomposition rate:

$$me = de = \frac{800 \text{ g} \cdot \text{m}^{-2}}{365 \text{ days}} = 2.2 \text{ g} \cdot \text{m}^{-2} \cdot \text{day}^{-1}.$$

Mean annual underground mortality rate per unit of surface = mean annual hypogeous decomposition rate:

$$mh = dh = 1.6 \text{ g} \cdot \text{m}^{-2} \cdot \text{day}^{-1}.$$

If the respective rates are calculated at a time when the processes reach their annual maxima, we obtain the following maximum values (epigeous and hypogeous combined):

$amax = 10 \text{ g} \cdot \text{m}^{-2} \cdot \text{day}^{-1}$, in the months of June and July

$mmax = 4 \text{ to } 5 \text{ g} \cdot \text{m}^{-2} \cdot \text{day}^{-1}$, in the months of June and July

$dmax = 10 \text{ g} \cdot \text{m}^{-2} \cdot \text{day}^{-1}$, in the months of May and June.

The above figures apply to the savannas of Calabozo. For the savannas of the llanos of Barinas, the mean values for the respective annual rates would be approximately 20 to 30% higher, owing to their greater production. But their maximum rates are probably lower because of the greater diversity of their phenorhythms, which allows growth, mortality, and probably also decomposition to spread more uniformly over the year.

For the savannas of Calabozo, the maximum rates of assimilation and mortality per unit of green weight and of decomposition per unit of dry weight are as follows:

$$a'max = 50 \text{ mg} \cdot \text{g}^{-1} \cdot \text{day}^{-1} = 5\% \text{ per day}$$

$$m'max = 20 \text{ mg} \cdot \text{g}^{-1} \cdot \text{day}^{-1} = 2\% \text{ per day}$$

$$d'max = 20 \text{ mg} \cdot \text{g}^{-1} \cdot \text{day}^{-1} = 2\% \text{ per day}.$$

Finally, the mean assimilation rate per assimilatory surface unit is as follows:

$$a'' = 3.8 \; g \cdot m^{-2} \cdot day^{-1}/(2.5 \; m^2 \cdot m^{-2}) = 1.5 \; g \cdot m^{-2} \cdot day^{-1}$$
$$= 0.15 \; mg \cdot cm^{-2} \cdot day^{-1}.$$

During the months of greatest assimilation, June and July, solar radiation in the llanos is close to its annual minimum due to the cloud cover and amounts to 10 to 12 Kcal $\cdot$ cm$^{-2}$ $\cdot$ month$^{-1}$, that is, between 333 and 400 cal $\cdot$ cm$^{-2}$ $\cdot$ day$^{-1}$. The assimilation rate during this period reaches the maximum value of 10 g $\cdot$ m$^{-2}$ $\cdot$ day$^{-1}$, which is equivalent to some 4 cal $\cdot$ cm$^{-2}$ $\cdot$ day$^{-1}$. This maximum rate implies a total efficiency of 1 to 1.3% with respect to the incident radiation energy, which, translated into utilizable photosynthetic energy, would represent an efficiency of double these values. We can therefore conclude that these tropical savannas with their many C4 grasses are capable of mantaining high productivity in their most active growth phase when water availability is not a limiting factor. But the factors limiting production do operate during the rest of the year, thereby reducing the mean annual efficiency of the assimilatory process to 0.3-0.4% of the total incident radiation energy.

Last of this series, table 8 summarizes various recycling turnover rates and times. It shows that the green biomass is renewed after approximately 3-5 months, while the total biomass renews itself every 18 months.

# Production data of other savannas and grasslands

The evaluations of production and productivity of tropical savannas in recent publications (Bourlière and Handley 1970;

**Table 8.** Turnover rates and turnover time of the vegetable matter in a savanna of *Trachypogon* of the central llanos.

| Turnover rates | Turnover time |
|---|---|
| Of V: $C_V = P_E/V_{max} = 1200/350 = 3.4$/year | $1/C_V = 0.3$ year $= 4$ months |
| Of S: $C_S = D_E/S_{max} = 1200/1000 = 1.2$/year | $1/C_S = 0.83$ year $= 10$ months |
| Of T: $C_T = P/T_{max} = 1800/1800 = 1$/year | $1/C_T = 1$ year |

Murphy, 1975) produce estimates that we consider to be very low. Most of these production estimates are calculated on the basis of the maximum peak of the aerial biomass. For instance, in the tabulation of the results published by Bourlière and Handley, the highest figures of annual production are those of Nye and Greenland (1960), describing a secondary savanna in Ghana with approximately 1500 mm of rainfall per year, and estimating on the basis of the peak of biomass an annual production of 870 g $\cdot$ m$^{-2}$ $\cdot$ year$^{-1}$. Hopkins (1968) studied a secondary savanna in Nigeria, with 1150 mm of rainfall as a mean during four years, and recorded an annual production of 680 g $\cdot$ m$^{-2}$. In this case the greatest biomass obtained in four years of measurements reached 800 g $\cdot$ m$^{-2}$ and the maximum value of the leaf area index (LAI) reached 4.56 m$^2$ $\cdot$ m$^{-2}$.

In Delhi, where rainfall amounts to only 800 mm and the season lasts three and a half months, Varshney (1972) obtained an aerial production of 798 g $\cdot$ m$^{-2}$ $\cdot$ year$^{-1}$ in a protected grassland in which *Heteropogon contortus* formed 90% of the biomass. He elaborated this aerial production by adding to the peaks of production of the dominant species that of the species of less importance. Even though the period of active growth is only 92 days, the mean rate of assimilation turned out to be 7.17 g $\cdot$ m$^{-2}$ $\cdot$ day$^{-1}$.

Ambasht et al. (1972) analyzed two protected parcels of grassland in the region of Varanasi, India, with a precipitation of 750 mm, concentrated in three and a half months. In a parcel dominated by *Heteropogon contortus* they determined monthly the total biomass (aerial and subterranean), separating it into the contribution of *Heteropogon* and that of the other species. They used the same method in a parcel dominated by *Dichanthium annulatum* where they separated the biomass of this species from the rest. In the first parcel, adding both biomass peaks produces a total annual production of 3225 g $\cdot$ m$^{-2}$ $\cdot$ year$^{-1}$; while if one adds the positive biomass one obtains a value of 3958 g $\cdot$ m$^{-2}$ $\cdot$ year$^{-1}$. For the *Dichanthium* grassland, the respective values calculated by both methods are 2024 and 2097 g $\cdot$ m$^{-2}$ $\cdot$ year$^{-1}$. Considering how brief the growing period is, the values for total biomass would indicate very high assimilation rates. Because the data make no discrimination between aerial and underground biomass, it is impossible to calculate hypogeous and epigeous productivity separately. In addition, since the peaks of epigeous and hypogeous growth seldom coincide in time, this method underestimates total production.

The biomass and production of the temperate grassland, the extratropical ecosystem with greatest similarities to the herbaceous stratum of the savannas of northern South America, were studied by several researchers from central Missouri. In this region, with 1000 mm of annual rainfall, the prairie reaches its maximum development, the so-called "tall-grass prairie." It forms a floristically rich community dominated by *Andropogon gerardi* and *A. scoparius*. Kucera and Ehrenreich (1962) and Kucera and Dahlman (1967) conducted a three-year study of this ecosystem, in conditions of fire protection, and determined a maximal aerial biomass of between 938 and 980 g $\cdot$ m$^{-2}$. The maximum of green biomass varied in the three years between 482 and 570 g $\cdot$ m$^{-2}$, while the maximum dry biomass varied in the same period between 410 and 474 g $\cdot$ m$^{-2}$. On the basis of these values Kucera and collaborators estimate the epigeous annual production at about 500 g $\cdot$ m$^{-2}$. In another study of the same prairie, Dahlman and Kucera (1965) measured the maximum hypogeous biomass at the end of the growing season, in the first 85 cm of soil and obtained a value of 1900 g $\cdot$ m$^{-2}$, indicating an underground production, estimated on the basis of increments in each of the three soil horizons, of 510 g $\cdot$ m$^{-2}$ $\cdot$ year$^{-1}$. That is, each year about 25% of the underground biomass is turned over. The total biomass of the ecosystem would be in the order of 2800–3000 g $\cdot$ m$^{-2}$, of which two-thirds correspond to the underground biomass; it can also be seen that the above- and below-ground production are approximately the same.

One of the most complete analyses on the structure and function of herbaceous ecosystems is the comparative study of ten grasslands in central and western United States carried out as part of the US/IBP Grassland Biome Program. It measured net primary production, compartmental transfers, and energy flows in grazed and ungrazed parcels (Sims and Singh, 1978a, b, c). For a comparison with tropical savannas the most useful will be the results of the tall-grass prairie site at Osage, Oklahoma (900 mm of precipitation), which seems to be a grass community that is closest to tropical grasslands, though obviously they are two quite distinct ecosystems. This tall-grass prairie, rather similar to that of the Missouri site, has *Andropogon scoparius* and *Sorghastrum nutans* as dominant grasses, growing on a rich mollisol. Intraseasonal changes in various primary producer compartments were followed for three years (1970 through 1972). Though important variations occurred during the three years, the average trends for the ungrazed prairie are about 300 g · m$^{-2}$ of live shoot biomass, about 400 g · m$^{-2}$ of recent dead, and 300 to 400 g · m$^{-2}$ of litter. Crown biomass varied erratically through the growing season, ranging from 200 to 500 g · m$^{-2}$, whereas root biomass (0 – 60 cm) attained between 680 and 1050 g · m$^{-2}$.

Net above-ground primary production was determined by the addition of peak live weights of individual species. It is therefore a conservative estimate because it did not include the material produced and decomposed before peak live weight of a given species occurs. The ungrazed tall-grass prairie's mean production for the three years was 346 g · m$^{-2}$ · year$^{-1}$, while below-ground net production, calculated by adding the significant positive increments in the crown and root biomass, could be estimated as 542 g · m$^{-2}$ · year$^{-1}$. That is, the total net primary production had annual values ranging from 847 to 933 g · m$^{-2}$. The figures obtained are thus comparable to those reported by other researchers, and again it is clear that in the tall-grass prairie of the temperate zone, as in the tropical grasslands, both below-ground biomass and production are higher than the respective above-ground values.

All the results that were discussed so far were based either on the peaks of biomass, whether total or epigeous, or on the total epigeous biomass divided into dominants and minor species whose maxima were added up, or on the positive increments in the biomass from which the minimum values of the respective biomasses were subtracted.

Another method of calculating productivity takes into account data from the processes of mortality and decomposition. The first researchers to calculate production in a grassland from this more rigorous point of view were Wiegert and Evans (1964). They measured decomposition utilizing the standard technique of litter bags, but also devised the technique of paired plots to determine the rate of mortality and from it the rate of epigeous decomposition. They took contiguous samples; in one of the plots they eliminated all the dry biomass (S) and after a certain time determined in both plots the increment in green and dry biomass (Iv and Is), obtaining in this manner an estimate of mortality, under the assumption that it would be equal to the negative increase in V. Adding the mortality to the measured increment of S one obtains the value of de.

However, in accordance with the productive model explained before (where R = reallocation) and which is applicable to any grassland, the mortality in any interval is:

$$M = -Iv - R + P.$$

Yet the method of Wiegert and Evans implies that during an interval of decreasing values of V there has been no reallocation to the hypogeous organs (R) nor production (P). Consequently one does not always obtain an acceptable estimation of M or of de, which is calculated as follows:

$$de = -Is \cdot t^{-1} + M \cdot t^{-1}$$

The measurable rates of decomposition for two Michigan secondary grasslands, as calculated by Wiegert and Evans in this manner, oscillated during the year between 1.3 and 13.6 mg $\cdot$ g$^{-1}$ $\cdot$ day$^{-1}$. The values we calculated for the savanna of *Trachypogon* oscillated during the year between 1.8 and 30 mg $\cdot$ g$^{-1}$ $\cdot$ day$^{-1}$. The epigeous production of a secondary grassland in southeast Michigan in which *Poa compressa* and *Aristida purpurascens* formed 90% of the biomass was as follows: 270 g $\cdot$ m$^{-2}$ counting only the total biomass increment without separating the species; 340 g $\cdot$ m$^{-2}$, (that is, 25% more), if the increments of each of the two dominants and the set of all the other species are added up; but when the mortality and decomposition rates were added, the estimate of the aerial annual production was increased between 60 and 80% above the first value. It is thus evident that failing to consider the totality of the production processes of the grassland leads to an underestimation of the aerial production in the order of 100%.

In Cañas, Costa Rica, Daubenmire (1972) studied the production of a secondary savanna that was burned each year, derived from a semideciduous tropical forest, and was entirely populated by *Hyparrhenia rufa*. The mean annual precipitation here is 1538 mm over seven months. The soil is deep and covers volcanic rock, so it is relatively rich in nutrients. Daubenmire calculated epigeous mortality and decomposition in a manner similar to the one used by Wiegert and Evans, eliminating dead matter in paired quadrats and measuring the increment of V and S. His production values were 515.4 g $\cdot$ m$^{-2}$ for the maximum of green biomass; 967 g $\cdot$ m$^{-2}$ for the maximum of aerial biomass, and 1378 g $\cdot$ m$^{-2}$ for the epigeous product. The criticism made of the work of Wiegert and Evans is also applicable here; that is, that although decomposition is measured, it is not taken into consideration that production, mortality, and decomposition occur simultaneously. However, the increment in the estimate of production with respect to the maximum epigeous biomass was only 42% in this case.

The primary production of the Lamto savannas, Ivory Coast, under 1300 mm of annual rainfall, was studied by Cesar and Menaut (1974) and Menaut and Cesar (1979). They considered seven savanna communities along a topographic and physiognomic gradient ranging from a pure grass savanna in a low-lying site over hydromorphic soil (hyperseasonal savanna in our terminology; see chapter 5) to a savanna-woodland of Andropogoneae, in a higher elevation, over ferruginous, well-drained soils. Maximum total above-ground biomass along this gradient ranged from 690 g $\cdot$ m$^{-2}$ in the grass savanna, to 1100 g $\cdot$ m$^{-2}$ in the Andropogoneae woodland. The respective figures obtained for maximum below-ground biomass (0–70 cm) for three of the seven savannas were (Cesar and Menaut, 1974) 2350, 1700, and 1680 g $\cdot$ m$^{-2}$ (the root-to-shoot ratio remains well above one). The aerial production of the herbaceous cover was calculated taking into account the peak of the biomass and a monthly decomposition rate. The values so obtained were 830, 1490, and 630 g $\cdot$ m$^{-2}$ $\cdot$ year$^{-1}$ for the three communities. That is an increment of between 20 to 35% with respect to the biomass maxima.

The below-ground production, calculated as the difference between *Hmax* and *Hmin*, is 1320, 1900, and 1220 g $\cdot$ m$^{-2}$ $\cdot$ year$^{-1}$ respectively. Adding these to the above-ground figures gives a total herbaceous production of 2150, 3390, and 2670 g $\cdot$ m$^{-2}$ $\cdot$ year$^{-1}$. These results, as the authors themselves say, are only rough estimates, since they do not include the decomposition in the soil. Finally, Menaut and Cesar estimated wood production through a dimension analysis of the four main woody species growing in the area. In the intermediate savanna, an open shrub community, wood production reached only 60 g $\cdot$ m$^{-2}$ $\cdot$ year$^{-1}$, whereas in the savanna woodland it attained 670 g $\cdot$ m$^{-2}$ $\cdot$ year$^{-1}$. That is, even in a savanna where trees reach their maximum height and density, the aerial wood production represents just about 25% of the total production attained by the grass layer.

Singh and Yadava (1974) conducted a very complete study of the seasonal variation of the biomass of a savanna protected from grazing in the north of India and used various methods to calculate the net primary productivity. Although the authors consider this site as tropical, its location at 29° 58' N does not qualify it as such;

the annual mean rainfall is 799 mm (713 during the measurement years), concentrated in three and a half months. It is a secondary vegetation, maintained by fire and grazing, that has been protected for twenty years. The floristic composition is relatively rich, with 46 species in 2 ha, of which 30 are annual plants. The most important species in terms of their relative contribution to the total biomass are *Panicum miliare* (16%), *Sorghum halepense* (6.7%), *Kochia indica* (18%), and *Chenopodium album* (18%); that is, two perennial grasses and two annual weeds. The researchers followed month by month the changes in the green and dry epigeous biomass ($E$, $V$, and $S$), as well as in the litter biomass and underground biomass to 30 cm of depth.

The peak of the maximum aerial biomass of all the vegetation reached 2350 g $\cdot$ m$^{-2}$, and if we subtract the initial epigeous biomass including the litter, in the order of 400 g $\cdot$ m$^{-2}$, we obtain an initial value for the epigeous production of around 1950 g $\cdot$ m$^{-2}$ $\cdot$ year$^{-1}$. In addition, Singh and Yadava (1974) calculated the epigeous production by four different methods:

1. Adding the positive differences of the total aerial biomass in successive samples, which then yielded a value of 2046 g $\cdot$ m$^{-2}$ $\cdot$ year$^{-1}$;

2. Adding up the maximum values reached by each species (28 separate species plus the rest of the minor species), which yielded a value of 2663.6 g $\cdot$ m$^{-2}$ $\cdot$ year$^{-1}$;

3. Adding for each species the positive increments in epigeous biomass that have taken place in successive harvests, which yielded a value of 3984.8 g $\cdot$ m$^{-2}$ $\cdot$ year$^{-1}$;

4. Adding the positive increments in total biomass plus the mortality calculated by the method of Wiegert and Evans (1964), which yielded a value of 2407 g $\cdot$ m$^{-2}$ $\cdot$ year$^{-1}$.

The hypogeous production was calculated adding the positive increments of the underground biomass; in this manner an annual value of 1131 g $\cdot$ m$^{-2}$ was obtained. The authors conclude that because methods (2) and (4) give equivalent results, they provide the best estimates of the epigeous production. Consequently the value of the net total primary production of the grassland would be some 3500 to 3800 g $\cdot$ m$^{-2}$ $\cdot$ year$^{-1}$. The hypogeous production estimate gives a low value, in part because the method used under-

**Table 9.** Values of the primary production obtained by different methods, in grasslands and savannas (g/m² per year).

| Author and country | Dominant vegetation | Precipitation (mm) | Increment of aerial biomass | Increment of aerial biomass by species | Aerial production plus decomposition | Increment of under-ground biomass | Total estimated production |
|---|---|---|---|---|---|---|---|
| Nye and Greenland Ghana | Secondary savanna, burned | 1500 | 870 | — | — | — | — |
| Hopkins Nigeria | Secondary savanna, burned | 1150 | 680 | — | — | — | — |
| Varshney India | Secondary grassland of *Heteropogon contortus*, not burned | 800 | 760 | 798 | — | — | — |
| Ambasht et al. India | Secondary grassland of *Dichantium annulatum* and *Heteropogon contortus*, not burned | 750 | — | — | — | — | — |
|  |  | — | 1359[a] | 2024[a] | — | — | 2024 |
|  |  | — | 2883[a] | 3225[a] | — | — | 3225 |

| Source | Location | Vegetation | | | | | | |
|---|---|---|---|---|---|---|---|---|
| Menaut and Cesar | Ivory Coast | Guinea savanna | 1300 | — | — | — | — | — |
| | | Grass savanna, burned | — | 690 | — | 830 | 1320 | 2150 |
| | | Intermediate shrub savanna, burned | — | 1110 | — | 1490 | 1900 | 3390 |
| | | *Andropogon* savanna woodland, burned | — | 690 | — | 630 | 1220 | 2670 |
| Daubenmire | Costa Rica | Secondary savanna of *Hyparrhenia rufa*, burned | 1538 | 967 | 967 | 1378 | — | — |
| Singh and Yadava | India | Secondary grassland of *Panicum miliare* plus weeds, not burned | 800 | 1950 | 2663 | 2407 | 1131 | 3500–3800 |
| Kucera et al. | U.S.A. | Tall-grass prairie of *Andropogon gerardi* and *A. scoparius*, not burned | 1000 | 500 | — | — | 510 | 1010 |
| Sims and Singh | U.S.A. | Tall-grass prairie of *A. scoparius* and *Sorghastrum nutans*, not burned | 900 | — | 346 | — | 542 | 847–933 |

[a] Total biomass (aerial plus underground).

estimates it, and also because we are dealing with a successional stage where there is a predominance of annual species of limited underground development. On the other hand, the value of the epigeous production obtained by method (3) is considered by these authors to be too high, owing to the difficulties of sampling species of very low biomass. In any case, the aerial production for this grassland is much higher than what was calculated for the Venezuelan savannas. The explanation possibly lies in the fact that communities rich in weedy species are more productive than well-established grasslands. Table 9 summarizes the results of the various studies.

# Summary

A few preliminary generalizations might be useful, if only to point out the big gaps that still exist in our understanding of the productive processes in tropical savannas. There is no method to estimate the hypogeous production of these ecosystems in a satisfactory manner. Without reliable data, it is not possible to understand fully the relationships between the hypogeous and epigeous biomass throughout the life cycle. On the other hand the ability to estimate epigeous production is improving, with better accounting for the contribution of the different species as well as for the rates of mortality and decomposition and their variations throughout the year. Even though the methods and principles used are not exempt from criticism, the results obtained by different researchers give us a first general view of the aerial production of various grassland and savanna ecosystems. But even these results are too fragmentary, both from a geographic point of view and from the perspective of the ecosystems themselves, to allow firm generalizations of general validity that can be extrapolated to any tropical savanna.

For the time being it might be best to limit ourselves to maintaining that the herbaceous production of the savannas is higher than what is suggested by most of the data in the literature, because data result from harvest methods that underestimate assimilation by not taking into account the totality of the productive processes that take place simultaneously. Very likely, the epigeous herbaceous production exceeds by 50% or 100% the values of the annual increment of the aerial biomass measured by successive harvests. The portion of the total assimilation allocated to the underground organs appears to vary in different types of savanna, being in some cases less than the epigeous production and slightly more in others. On the other hand, even in savannas with the highest cover and density of trees, the aerial production of this stratum does not attain more than 25% of the herbaceous stratum.

Some secondary herbaceous communities in tropical zones of India and Central America have higher total production values than natural grasslands and savannas. In India, in particular, these communities can be classified as highly productive. In tropical America the secondary savannas of introduced African species, like *Hyparrhenia rufa*, are more productive than the primary savannas dominated by native species. However, a comparison of African with American savannas does not seem to disclose greater productivity in the former. It is thus likely that also in Africa the secondary grasslands or savannas are more productive than the natural communities formed by shorter species. In any case, this phenomenon of greater productivity of seral communities is in line with the known facts about the high rates of growth of pioneering and weedy species.

Finally, the data point out that the maximum assimilation rates that can be reached during a few weeks can be very high, comparing favorably with rates for other tropical crops. This happens in the most active growth phase of the photosynthetic apparatus, which in different savannas coincides with different stages of the wet season. Nevertheless, during most of the year, other factors severely limit production in these ecosystems, significantly reducing the values of the mean annual assimilation rates. Further on we will discuss the role that water and mineral nutrients play in limiting production.

# 5

# Water economy

Although the American savannas, in contrast to African and Australian savannas, are exclusively humid, tropical ecosystems (Sarmiento and Monasterio, 1975), water represents one of the environmental elements that most influence the structural and functional characteristics of the system. Disposable water is undoubtedly the principal factor that regulates the establishment and survivability of the species and savanna formations in each habitat.

To understand the essential role of water it is necessary to reemphasize that savannas are intrinsically seasonal systems; that is, their fundamental ecological processes develop in two contrasting rhythms during each annual cycle, passing from a period of intense and diversified activity to one of more or less prolonged dormancy. This intrinsic seasonality is clearly expressed in the structural and functional changes of the ecosystem throughout the year, with all the complexity of the successive or simultaneous development of different phenophases that characterize the evolutionary strategies adopted by each group of species of similar phenological behavior. In a similar manner, these changes constitute an essential part of the rhythmicity shown by the productive processes (see chapter 4).

It is also necessary to point out that the two essential but opposite stages of growth and rest, which integrate the annual cycle of all savannas, are always synchronized with an external seasonality where water plays a controlling role. At the same time, one must not fall in the trap of assuming a simple and direct causal relation of seasonal water availability as the principal cause, and the responses of the ecosystem as the immediate consequence. It is obvious that

the savanna species have evolved through a prolonged evolutionary process in response to a constellation of external stimuli and
selective forces of the physical and biotic environment, which have
not always been constant in time and do not have the same impact
on all the structural and floristic components of the ecosystem.

The American savannas are essentially seasonal systems whose
function is adjusted to the most restrictive rhythm of the environment they occupy. The problems of their water economy are quite
different from those of a system that is permanently subjected to
one or the other of the two extreme humidity conditions, such as
arid formations under permanent conditions of water deficit, or
marshy formations with too much water in the soil. For this reason
savannas cannot be described, other than in a very crude way, as
"dry" or "wet," since what really must be considered is the inter-
play between the combination of various hydric regimes during
successive stages of the annual cycle. A schematic description of
some life cycle alternatives follows.

1. A prolonged wet season, which causes problems of excess
water in the horizon exploited by the dominant grasses, followed
by a dry period when these same species are faced with a water
shortage. When this sort of alternation of two very contrasting
seasons takes place, we call the corresponding ecosystem a hyperseasonal savanna.

2. A prolonged season with excess soil water, followed by a
season with sufficient soil water in the upper soil layers (if there is a
shortage, it is only partial). This is the ecosystem that we will call
marshy savanna in accordance with the popular Venezuelan nomenclature *(sabana de estero)*.

3. A season with sufficient moisture in the upper soil layers, but
always below saturation levels, alternating with another season
with marked soil water deficit. This describes the situation of the
systems that we call seasonal savannas.

4. A long wet season but without an excess of moisture, followed
by one with partial deficits, whether because of its short span or
because it does not extend to the usable profile by grasses. We will
call this a semiseasonal savanna, since it is the one that suffers least
from the impact of the oscillations of the hydric regime.

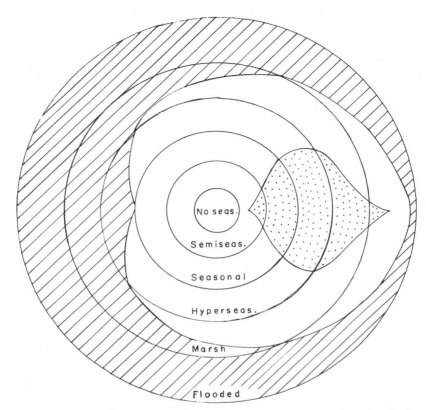

**Figure 27** Diagrammatical representation of different water regimes in seasonal ecosystems. Each circle corresponds to the annual cycle of an ecosystem; the *hatched* area represents the period of water excess, the *white* area that of normal water availability, and the *dotted* area the period of water deficit. From the center to the periphery the formations are: an ecosystem without a water-induced season; the semiseasonal savanna; the seasonal savanna; the hyperseasonal savanna; the marsh; and a wetland ecosystem.

The various types of extrinsic or environmental seasonality are depicted in figure 27, where each concentric ring represents a different hydric type of habitat-vegetation, from a nonseasonal ecosystem in the center, with water available all the time, to a marshy formation at the periphery, with continuous saturation conditions. The respective circumferences indicate the type of yearly cycle for each ecosystem, with separate phases of saturation, of disposable water, and of moisture deficit. The figure then shows how the length of the phase of excessive water increases from the center to the periphery, while the period of water deficit is steady in the central ecosystems, diminishing to the periphery.

This scheme of qualitative differentiation of various ecological types of tropical savannas on the basis of conditions determined by yearly variation in water availability does not pretend to make an exact quantification either of the length or of the intensity of the periods that are considered "dry," "normal," or "moist" (more precisely "hypermoist"); neither have the external factors that give rise to them been pointed out, since these can be climatic, topographic, geomorphological, or edaphic. Despite these caveats, this distinction seems to be useful for savannas that are dissimilar in their fundamental structural and functional characteristics.

The example that most closely approximates the perhumid extreme is that of the marsh ecosystem. During the wettest period of the year it must support conditions of moisture saturation in all the soil profile utilized by the grasses, and in addition it suffers a prolonged period of waterlogging with standing water that can surpass a meter. Even during the dryest period the phreatic level remains high, leaving at most only the upper centimeters of soil under hydric stress, and then only for short intervals. The moisture conditions approximate those of swamps, which are inundated or permanently saturated most of the year.

The hyperseasonal savanna has two environmental limitations that are equally strong but opposite: in one season of the year the soil is saturated and even covered with a few centimeters of water, while in the other the soil horizon utilized by the grasses dries out completely. The semiseasonal savanna suffers only a relatively short period of drought. The seasonal savanna, on the other hand, is never subjected to prolonged conditions of soil saturation in the upper soil layers, but these soil layers do dry out completely for several months. In summary, marshes and semiseasonal and seasonal savannas are all biseasonal systems, since their life cycle is exposed to two moisture regimes: saturated and normal in the first case, normal and partially dry in the second, and normal and dry in the last instance. On the other hand, the hyperseasonal savannas are four-season systems, since in each life cycle they go through four periods: first one of excess, then a normal but drying period, then a third period of water deficit, and then again a normal but increasingly wet period. This classification helps to distinguish physiognomically similar but fundamentally different ecological types. Unfortunately, the duration and intensity of each moisture period for each of the different savanna types have not yet been quantified.

# The seasonal variations of soil moisture

The Rawitscher "school" in São Paulo pioneered research in transpiration and water balance in neotropical savannas, publishing a copious amount of data on numerous species of trees, shrubs, and grasses of the cerrados of central and southern Brazil (Ferri, 1944, 1955; Rachid, 1947; Ferri and Coutinho, 1958). Rawitscher himself was the first person to analyze the relation between transpirational losses and the variations in soil water content (Rawitscher et al., 1943; Rawitscher, 1948). Working in a cerrado in Emas, state of São Paulo, which has a savanna-type climate *(Aw)* with 1400 mm of rainfall over 7–8 months, Rawitscher followed the moisture content of the soil down to the phreatic level, throughout the dry season. The savannas of Emas are rich in woody species, and occupy latosolic soils developed on a very deep alterite of about 20 m. In this part of the Brazilian planalto, the phreatic level is found throughout the year between 17 and 18 m.

Rawitscher found that the moisture content at each soil level diminished during the dry months from the surface down to 2–3 m, but below this level the oscillations were dampened and disappeared completely at lower levels. He also confirmed previous research indicating that in some cerrados the deep-rooted woody species do not curtail their transpiration during the dry season, and their stomatal movements are rather slow. Neither do some grasses, such as the dominant species *Echinolaena inflexa*, control transpiration, but instead they dry out almost completely during the dry season. Rawitscher concluded that water cannot be a limiting factor in the development of deep-rooted woody species; because these species continue to transpire freely during the dry season, there must always be enough disposable water in the lower levels of the soil.

The water economy of the savannas of Rupununi, Guyana, is one of the central themes of the integral study of these systems performed by Eden (1964). The savanna climate *(Aw)* would have an annual rainfall of slightly over 1600 mm and additional monthly rains below 50 mm during four to five months. Eden measured the oscillations in the soil water content throughout most of the year at five sites that can be arranged in a topographical sequence along a moisture gradient, starting with a well-drained tropical lateritic soil (latosol), to a moderately drained one, to an imperfectly drained groundwater lateritic soil, then to a poorly drained one (these four with grassy savanna vegetation), and on to the last, a very poorly drained humic glei with swampy vegetation. At each site Eden followed the oscillation of the phreatic level and the percentage of soil moisture on two levels: 0 to 30 cm, where the root mass is concentrated, and 30 to 60 cm, where most of the remaining hypogeous biomass is found.

The results of Eden's work are summarized in table 10. The months during which soil moisture is above field capacity are classified as months with excess water at a certain soil level; that is, when there is gravitational water that has not drained. Months with a water deficit are those during which soil moisture is constantly below the wilting point. Table 10 indicates that the first site with a well-drained lateritic soil corresponds to what we have called semiseasonal savannas, with no excess or deficit of water in all of the soil horizons exploited by the roots. For approximately two months there is a water deficit in the 0 – 30 cm level, but never in the entire profile. This is also the only savanna in the entire gradient that possesses a tree stratum. The second site, with a moderately well-drained soil and a high phreatic table during one period, is also a semiseasonal savanna, but in contrast to the first one it has a water surplus in the 30 – 60 cm level, as a result of capillarity when the water table raises. The third and fourth communities with hydromorphic soils correspond in our classification to hyperseasonal savannas, since their cycle includes a period of 3 to 5 months of an excess of water in the entire profile and a period of deficit, at least in the upper horizons, which contain most of the hypogeous biomass (over 80%). The last community corresponds to a marsh where in the dry season the water table descends only to the 20 cm level. The work of Eden is the first quantitative presentation of the annual variation of the moisture conditions along a humidity gradient in a savanna, though unfortunately the seasonal savanna is missing from the sequence.

Askew et al. (1971) analyzed the role of water and nutrients in determining the limit between savannas and rain forests in the region of the Serra do Roncador (Mato Grosso, Brazil). The regional climate is *Aw* (savanna type), but very close to the category of the

**Table 10.** Variation in soil humidity in the savannas of Rupununi, Guyana (data from Eden, 1964).

| Type of soil | Min. depth of water table (m) | Max. depth of water table (m) | Months in yearly cycle | | | | |
| --- | --- | --- | --- | --- | --- | --- | --- |
| | | | Excess water at 0–30 cm | Excess water at 30–60 cm | Water shortage at 0–30 cm | Water shortage at 30–60 cm | |
| Well-drained latosol | 1.5 | 4.5 | 0 | 0 | 2 | 0 | |
| Moderately drained latosol | 1 | 5 | 0 | 3 | 2 | 0 | |
| Hydromorphic latosol, imperfectly drained | 0 | 4 | 3 | 5 | 7 | 0 | |
| Hydromorphic latosol, poorly drained | 0 | 2 | 5 | 6 | 2 | 0 | |
| Humic glei, badly drained | 0 | 0.2 | 7 | 7 | 0 | 0 | |

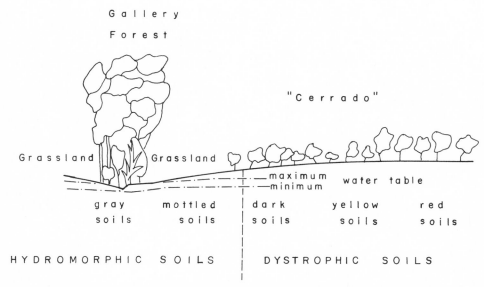

**Figure 28** Schematic profile from the cerrado to the gallery forest in the Serra do Roncador (Mato Grosso, Brazil), showing the edaphic and vegetational variation. (From Askew et al., 1971.)

monsonic climate *Am*; rainfall during the year was 1370 mm. The researchers followed the annual variation of soil moisture along four transects that reached from the cerrado vegetation (woody savanna) over oxisols, through campo vegetation (grassy savanna) over groundwater laterites (tropaquents), to the gallery forest in the bottom of the talwegs over poorly drained groundwater laterites (hydraquents; fig. 28). In the transects the change in the water table was followed, and samples were taken at three depths for gravimetric determinations of soil moisture: 4–8 cm, 45–55 cm, and 95–105 cm.

These results show that the water table in the gallery forest is always close to the surface and only seldom descends below 50 cm. The campos seldom flood though the water table fluctuates during the year between 50 cm and 2 m. In the area of the woody savannas (cerrado) the water table is very deep, below 30 m during the dry season; only in low-lying areas near the campos is it closer to the surface, oscillating between a minimum of 127 cm during the wet season and a maximum of 345 cm during the dry season. This information is sufficient to differentiate the three kinds of systems.

The gravimetric determinations indicate that areas with grassy savanna, except for the very low-lying ones near the gallery forests, have no usable water in the whole profile during the dry season, although the effect is less marked at 1 m than at 50 cm. During the rainy season, however, all the profile has usable moisture, although the soil never becomes waterlogged. The woody savanna sites have sufficient moisture in all three analyzed depths in the wet season; in the dry season their desiccation is more marked than in the grassy savanna. Paradoxically, the water content of cerrado soils increases with slope, a fact which the authors of this work are unable to explain satisfactorily.

According to the data of Askew and his colleagues, the grassy savannas of this region correspond to the savannas that we have called semiseasonal, because the drought is not too pronounced, and although the soil dries to a depth of a meter, the proximity of the water table suggests that between 1 and 2 m there is water available throughout the year. The cerrados on the other hand would be seasonal savannas, with water available only in very deep soil layers that cannot be reached by the roots of the grasses. These results in Mato Grosso confirm Rawitscher's observation that even if the soil around the roots of the grasses dries out, it retains enough water at the lower depths to fill the needs of the deep-rooted trees.

Foldats and Rutkis (1975) analyzed the variation in the water table on several sites of a 300 ha parcel, in the course of their phenological study of two tree species (*Curatella americana* and *Byrsonima crassifolia*) in the savanna of the Biological Station in Calabozo. For almost a year they followed the oscillations of the water table at six sites with savanna or with semideciduous forest, and for three years at a site with rain forest.

All the curves show the same tendency, regardless of whether they correspond to savanna or rain forest. Toward the middle of the rainy season (September), the water table increases up to levels of 1 to 3 m deep, and then gradually decreases until it reaches its lowest level in May or June (3 to 5 m). In very dry years, like one of the three in which the measurements were conducted, it can descend to 7 m. That is, practically from the moment the rain ceases the phreatic water is out of the reach of the herbaceous species, but it will remain available for the woody species throughout the year.

The studies of Foldats and Rutkis on transpiration confirm what other authors in Brazil and Venezuela had pointed out; the two woody species actively transpire during the whole dry season, and at least *Byrsonima* shows no stomatal control even during the times of greatest solar radiation. It can therefore be assumed that both tree species with their deep roots utilized the water resources in the lower soil levels, and that they can reach the water table most of the time.

To complement their studies on production in the savannas of *Trachypogon* at the Biological Station, San José and Medina (1975) followed the soil moisture variations in the burned parcel as well as in the protected one. During the year of the study rainfall was exceptionally high for the locality: 1839 mm. The water content was determined with a Bouyoucus probe at depths of 5, 10, 20, 40, and 70 cm, with daily readings throughout the year. Gravimetric computations were conducted monthly in order to calibrate the resistance curve.

The measurements of the water content in the soil column from the surface to 70 cm, showed that moisture stayed at $-15$ bars during most of the period between December and the middle of April, although the exceptional rains in February increased the soil moisture values during that month. The soil remained wet for about one month following the end of the rainy season (November), and after that it remained ecologically dry for the rest of the dry season, except for the unusual rains in February.

The measurements of the moisture values of the different horizons indicate that they all follow the rains, but the more superficial ones are also the more sensitive ones, wetting and drying quickly at the beginning and end of the rainy season and responding also to isolated showers during the dry season. The deepest measured horizon (70 cm) does not respond to the slight rains of the dry season; it stays wetter for a longer period after the rainy season (until the end of January, or two and a half months later) and dries less thoroughly, staying slightly above the $-15$ bar level until the beginning of the rains. The research of Foldats and Rutkis (1975) established that the water table in these savannas descends during the dry season to between 3 and 5 m, while during the wet season it ascends to close to 1 m, so that throughout the year there is sufficient available moisture at depths below those measured by San José and Medina.

San José and Medina also measured the transpiration of the three dominant species of grasses at several instances during the wet and dry season, correlating it with evaporation in tank A. On the basis of

**Table 11.** Vegetation and soil characteristics of six ecosystems in the western llanos.

| Soil series | Accumulation and relief forms | Vegetation: dominant species | Depth (cm) | Texture |
|---|---|---|---|---|
| Barinas | Q2 Alluvial core | Seasonal savanna *Axonopus purpusii* *Leptocoryphium lanat.* *Trachypogon vestitus* | 0–13 13–40 40–100 100–130 130–200 | Fa FAa FAa Fa FA |
| Garza | Q2 Compound alluvial core | Seasonal savanna *Paspalum plicatulum* *Axonopus purpusii* *Trachypogon plumosus* | 0–8 8–23 23–40 40–75 75–105 105–152 152–200 | Fa Fa FAa FA FA FAa Fa |
| Jaboncillo | Q2 Decantation valleys | Hyperseasonal savanna *Sorghastrum parvifl.* *Andropogon selloanus* *Leersia hexandra* | 0–12 12–28 28–40 40–50 50–100 100–130 130–160 160–200 | Fa FL F A A-Fa FA FA Fa |
| Boconoito (Sabana) | Q3 Alluvial fan | Seasonal savanna *Leptocoryphium lanat.* *Elyonurus adustus* | 0–10 10–56 56–74 74–120 120–165 165–200 | Fa FAa FAa FAa FAa Aa |
| Boconoito (Bosque) | Q3 Alluvial fan | Semideciduous forest *Acrocomia lasiopatha* *Genipa caruto* *Xylopia aromatica* *Trichilia spondioides* | 0–12 12–42 42–72 72–190 190–200 | Fa FAa FAa FAa Aa |
| Torunos | Q1 Loamy alluvial plain | Gallery forest *Anacardium excelsum* *Pithecellobium saman* *Attalea maracaibensis* | 0–16 16–30 30–50 50–60 60–90 90–112 112–167 167–200 200–250 | FAL AL FAL AL FAL FAL FAL FL F |

| Permeability | pH | Organic carbon (%) | Total nitrogen (%) | C.E.C. (meq/100 g) | Saturation (%) | Taxonomic classification |
|---|---|---|---|---|---|---|
| Mod-mod. fast | 5.4 | 0.96 | 0.06 | 3.25 | 51.7 | Oxic Paleoustalf |
| Moderate | 5.0 | 0.54 | 0.04 | 3.00 | 21.7 | |
| Mod. slow | 5.2 | 0.15 | 0.03 | 3.12 | 20.2 | |
| Mod. slow | 5.4 | 0.18 | 0.02 | 4.12 | 30.8 | |
| Moderate | 5.6 | 0.10 | 0.02 | 6.62 | 41.8 | |
| Fast | 5.2 | 1.18 | 0.08 | 3.37 | 43.0 | Oxic Paleoustalf |
| Mod. fast | 4.4 | 0.41 | 0.04 | 2.62 | 18.7 | |
| Moderate | 5.0 | 0.36 | 0.05 | 3.87 | 13.7 | |
| Moderate | 5.3 | 0.25 | 0.04 | 4.50 | 23.7 | |
| Moderate | 5.6 | 0.10 | 0.03 | 4.25 | 50.5 | |
| Moderate | 6.5 | 0.02 | 0.02 | 5.00 | 83.4 | |
| Mod-Mod. fast | 6.0 | 0.02 | 0.03 | 8.00 | 92.2 | |
| Moderate | 5.0 | 1.03 | 0.08 | 2.25 | 32.9 | Typic Tropaqualf |
| Slow | 5.2 | 0.23 | 0.03 | 1.00 | 29.0 | |
| Slow | 6.3 | 0.20 | 0.04 | 1.50 | 58.0 | |
| Slow | 6.8 | 0.31 | 0.04 | 8.20 | 86.7 | |
| Slow | 8.2 | 0.10 | 0.03 | 11.62 | 100 | |
| Slow | 7.7 | 0.10 | 0.03 | 12.00 | 100 | |
| Mod. slow | 7.5 | — | 0.02 | 10.00 | 100 | |
| Mod. fast | 7.1 | — | 0.01 | 4.50 | 100 | |
| Fast | 4.0 | 1.10 | 0.07 | 2.8 | 29.9 | Ultic Haplustalf |
| Moderate | 4.3 | 0.74 | 0.04 | 2.8 | 5.3 | |
| Moderate | 4.1 | 0.51 | 0.04 | 2.4 | 6.2 | |
| Moderate | 4.2 | 0.44 | 0.02 | 2.4 | 5.8 | |
| Moderate | 4.2 | 0.24 | 0.02 | 2.1 | 6.6 | |
| Moderate | 4.4 | 0.19 | 0.02 | 2.3 | 6.5 | |
| Fast | 4.3 | 0.81 | 0.05 | 1.9 | 14.2 | Ultic Haplustalf |
| Moderate | 4.3 | 0.49 | 0.03 | 1.9 | 11.0 | |
| Moderate | 4.2 | 0.46 | 0.03 | 2.1 | 10.0 | |
| Moderate | 4.3 | 0.24 | 0.02 | 2.0 | 8.5 | |
| Moderate | 4.6 | 0.12 | 0.02 | 2.0 | 14.0 | |
| Moderate | 6.3 | 3.49 | 0.25 | 21.0 | 100 | Fluventic Ustropent |
| Moderate | 6.0 | 1.54 | 0.11 | 14.9 | 100 | |
| Moderate | 6.3 | 0.63 | 0.07 | 11.0 | 100 | |
| Moderate | 6.4 | 0.54 | 0.06 | 11.6 | 100 | |
| Moderate | 6.4 | 0.89 | 0.07 | 11.5 | 100 | |
| Mod. fast | 7.3 | 0.39 | 0.05 | 10.2 | 100 | |
| Moderate | 7.9 | 0.24 | 0.03 | 7.6 | 100 | |
| Mod. slow | 7.7 | 0.17 | 0.03 | 7.8 | 100 | |
| Mod. slow | 6.4 | 0.17 | 0.02 | 5.8 | 100 | |

the evaporation and the leaf area index they calculated the daily and monthly transpiration of the savanna, obtaining values of 1440 and 1105 mm/year in the burned and protected parcels, respectively. These calculations assume that the grasses do not control transpirational loss and that they continue to transpire until they wilt when they run out of soil water.

Our own results from the llanos occidentales of Venezuela (Sarmiento and Vera, 1977) will round out this survey. During a per ꞏd of 18 months, which included two complete dry seasons, we followed the variations in the water content of the soil of six ecosystems, four savannas and two tropical forests, all situated close to each other in the region of the llanos of Barinas (see fig. 2).

Three of the savannas were at Hato Caroni, with 1017 mm of rainfall in the measurement year. One is a savanna of *Axonopus purpusii, Leptocoryphium lanatum,* and *Trachypogon vestitus* (Barinas); another has as dominants *Paspalum plicatulum, Axonopus purpusii,* and *Trachypogon plumosus* (Garza); the last is dominated by *Sorghastrum parviflorum, Andropogon selloanus,* and *Leersia hexandra* (Jaboncillo). The distribution of the hypogeous biomass in the profile was already shown in figure 9; table 11 describes the most important ecological characteristics of the soils.

At another locality, Corozo-Palmitas, with 1170 mm of rainfall in the year of measurement, we followed the moisture regime of two adjacent formations on the same soil: an open savanna and a semideciduous forest. The savanna's dominants of the woody stratum are *Byrsonima crassifolia, B. coccolobaefolia,* and *Bowdichia virgiloides,* while the dominants of the herbaceous stratum are *Leptocoryphium lanatum, Elyonurus adustus,* and *Trachypogon plumosus* (savanna of Boconoito). The forest canopy is some 15 m in height, apparently seral in character, dominated by *Acrocomia lasiopatha, Genipa caruto, Xylopia aromatica,* and *Trichilia spondioides.*

Soil moisture was also measured in a well-preserved parcel in the gallery forest of the river Santo Domingo, with a yearly rainfall of 1350 mm. This is a seasonal forest, 30 m in height, dominated by *Pithecelobium saman* and *Anacardium excelsum,* with the palm *Attalea maracaibensis* as the principal component of the understory stratum.

Soil moisture was measured with an ohmmeter and sensors of fiberglass permanently placed at 20, 50, 100, 150, and 200 cm, in three places in each ecosystem. In addition periodic gravimetric

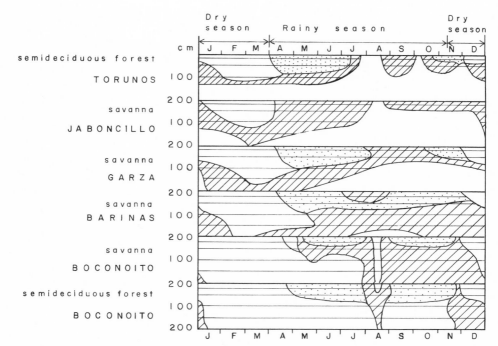

**Figure 29** Seasonal variations in soil moisture in six ecosystems of the llanos of Barinas, following periodic determinations at 20, 50, 100, 150 and 200 cm. The region in *black* represents the stage of water saturation, the *diagonal hatching* indicates that there is water available; the *dotted* area corresponds to the intermittently dry soil; the zone in *white* indicates when the soil is always below the point of permanent wilting. (From Sarmiento and Vera, 1977.)

determinations for control were performed. The results are summarized in table 12 and figure 29. As before, excess of water means that the soil water potential remains constantly over field capacity $(-\frac{1}{3}$ bar), and water deficit means that the soil water potential is permanently below the wilting point ($-15$ bars).

In the gallery forest the water table oscillates throughout the year between a depth of 2 and 3 m and the elevation of water by capillarity reaches the 1.50 m level, which consequently is saturated throughout the year. The two upper levels (20 and 50 cm) stay wet for a short time during the rainy season, but dry out completely during the dry season, while the 1 m level has six months of water excess and three months of shortage. Throughout the year the trees of this forest have access to a source of water either in the water table or from rainfall, while at the same time there is never an excess of soil moisture. Trees can use the water from the upper horizons during the rainy season, and during the dry season, the layers from 1 m down to the area of the water table. From the point of view of the water economy this ecosystem does not have any major limitations.

The situation of the semideciduous forest is totally different. Here the water table is very deep, there is never an excess of water in the two upper meters of soil, no usable soil moisture on the surface layers during all of the dry season, and none for seven months of the year at lower depths. At this site usable water is provided entirely by rainfall, and soil wetting takes place exclusively from above. The trees have only two alternatives: to eliminate water losses during the dry season by total loss of foliage, or to limit dehydration through appropriate morphological mechanisms, as is the case with evergreen species, all of which are highly scleromorphic.

Table 12 shows that all four savannas suffer a complete lack of water, at least during the three to four months of the dry season, in the entire profile used by grasses. But the savannas differ from each other in the rainy season. For example, the savanna of Boconoito (seasonal) never has an excess of water at any soil level; Barinas (seasonal) has an excess only in the deepest horizon (2 m), which is hardly used by the roots of the grasses; Garza (seasonal), on the other hand, shows a prolonged period with an excess of water, from 1 m down, due to a hanging water table, but these saturation conditions do not reach the levels where the maximum root concentration is found; finally, Jaboncillo has a prolonged period of water saturation from 40 cm down and a period of excess and even saturation in the surface levels, and can therefore be classified as a hyperseasonal savanna.

Figure 29 shows the different moisture periods in each soil level for the six ecosystems. The dry period is viewed as having two aspects. The first is the aspect of the climatically dry season, when a measurement below the wilting point is taken as an indication that this is the condition of that horizon throughout the measurement period. The second situation occurs during the rainy season when a water potential below − 15 bars is recorded; this is merely temporary and means that the vegetation had utilized all the available water and will remain short of water only until the next rainfall.

Another study provides additional information on the variation

**Table 12.** Maximum number of months in which a soil level may be saturated or dry in some ecosystems of the western llanos of Venezuela (data from Sarmiento and Vera, 1977).

| Soil depth (cm) | Gallery forest on fluventic ustropent, moderately well drained (wet dry) | | Semideciduous forest on ultic haplustalf, well drained (wet dry) | | Open savanna of Byrsonima crassifolia, Leptocoryphium lanatum on ultic haplustalf, well drained (wet dry) | | Grassy savanna of Axonopus purpusii on oxic paleoustalf, with deep hydromorphy (wet dry) | | Grassy savanna of Paspalum plicatulum on oxic paleoustalf, with deep hydromorphy (wet dry) | | Grassy savanna of Sorghastrum parviflorum on typic tropaqualf, imperfectly drained (wet dry) | |
|---|---|---|---|---|---|---|---|---|---|---|---|---|
| 20 | 1 | 4 | 0 | 4 | 0 | 4 | 0 | 4 | 0 | 4 | 1 | 3 |
| 50 | 1 | 4 | 0 | 4 | 0 | 4 | 0 | 4 | 0 | 4 | 4 | 3 |
| 100 | 6 | 3 | 0 | 4 | 0 | 4 | 0 | 4 | 5 | 4 | 5 | 2 |
| 150 | 12 | 0 | 0 | 7 | 0 | 7 | 1 | 4 | 6 | 3 | 8 | 0 |
| 250 | 12 | 0 | 0 | 7 | 0 | 7 | 3 | 4 | 6 | 2 | 10 | 0 |

of soil water during an annual cycle and the water balance of the species of a seasonal savanna in the llanos of Barinas. Goldstein et al. (1982) measured the water content and water potential at 10, 40, and 70 cm in the soil and determined the daily course of foliar conductance, transpiration rate, and leaf water potential of four woody and three grass species. Preliminary results indicate that during the dry season the trees lose large amounts of water through transpiration, even though they may be able to control these losses to some extent through partial stomatal closing. Leaf water potentials reached values of $-15$ to $-20$ bars during the times of greatest transpirational loss, while evening and dawn water potentials were of $-2$ to $-3$ bars. These values were obtained when the soil water potential in the 10 and 40 cm levels reached $-15$ bars, but at 70 cm they were less negative. Transpiration of perennial grasses was comparatively lower, even though their midday leaf water potential was $-20$ bars, and their dawn potential oscillated between $-5$ and $-10$ bars. That is, even the perennial grasses dispose of limited water resources in the soil: that translates into some rather negative water potentials in the plant.

This survey of the soil moisture studies should end with the caution that statements such as "disposable water," "permanent wilting point," and "field capacity" are schematic first approximations, not final determinations of the real water dynamic of the soil-plant-atmosphere system. One must not take these expressions literally. Thus the concept of "permanent wilting point" has almost no relation to the real wilting point of an individual plant; furthermore, the amount of disposable or usable water for a particular species in one community does not necessarily have to be the same as what that species can extract from a different site under the same conditions.

# Elements for a discussion of the water balance of tropical savannas

Our knowledge of the variables that determine the water balance is still too fragmentary to establish in a precise and rigorous manner the annual water balance of a savanna. This discussion will therefore reason through several aspects of the water economy, utilizing, as in the chapter on productivity, the methodological procedure of presenting the simplest model applicable to the system. The model will help to visualize the set of intervening relationships,

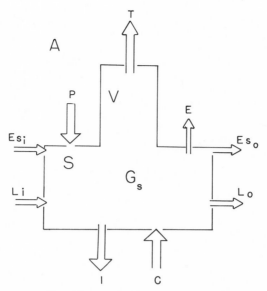

**Figure 30**  The water balance model in a savanna ecosystem. The vegetation-soil compartment (*V* and *S*) is continuous and separated from the atmosphere *A*. *Gs* corresponds to the amount of useful water in the soil. The inputs to the *VS* subsystem are: precipitation *(P)*, capillary movement *(C)*, surface drainage from high areas *(Es$_i$)*, lateral subsurface infiltration *(L$_i$)*. The losses from the system are: evaporation from the soil *(E)*, transpiration *(T)*, deep infiltration *(I)*, and the losses from surface drainage *(Es$_0$)* and lateral infiltration *(L$_0$)*.

some of which can be quantified while others can only be inferred in an approximate manner.

As far as the movement of water is concerned, the soil-plant-atmosphere system can be considered as a continuum (Philip, 1966), both for the individual plant and for the entire vegetation. A parcel of land and the air above it thus form a coherent system subdivided into two subsystems (figure 30): one of them, well delimited physically, is the *VS* (vegetation-soil); the other, with less precisely defined limits, is *A* (atmosphere).

The influx of water *(G)* into the subsystem *VS* can take place through four different channels, as follows:

Precipitation *(P)* coming from *A* that reaches the *A-VS* interphase.

Surface runoff *(Es)* that originates in another parcel of the same or a different ecosystem and infiltrates this one.

Lateral influx *(L)* that arrives into the system below the surface from water infiltrated at higher elevation sites.

Water that ascends through capillarity *(C)* from a nearby water table.

The total input of water into the *VS* system during a set period is:

$$G(in) = P + Es(in) + L(in) + C$$

On the other hand the components of the output of water from the *VS* system are:

Evaporation *(E)* from the soil surface and vegetation (interception).

Transpiration *(T)* of the vegetation into the atmosphere.

Surface runoff *(Es)* that leaves the parcel to lower sites.

Lateral losses *(L)* due to subsurface drainage to lower-lying places.

Infiltration or internal drainage *(I)* to lower soil levels not considered as part of the system.

After a given time interval the water losses are:

$$G(out) = E + T + Es(out) + L(out) + I$$

The water accumulated in the *VS* system will be called *Gs*. This includes water retained by the plant biomass of the savanna, which is at least two orders of magnitude less than what is accumulated in the soil profile at any one time. For a given period we will then have a variation of water content that is:

$$Gs = G(in) - G(out)$$

This model will be applied to one of the seasonal savannas that we have studied in the llanos of Barinas. The savanna of Boconoito has a tree density of 100 individuals per ha, with *Byrsonima crassifolia* as dominant species. The herbaceous stratum is dominated by *Leptocoryphium lanatum* and *Elyonurus adustus*. In this ecosystem not all the variables of the model are significant; several can be disregarded in a first approximation, and others discarded completely. Since we are dealing with a flat savanna, with a very slight slope and a high plant cover during most of the year (except just after fire), we will not take surface runoff into consideration. Likewise, in view of the type of relief of the terrain and the depth of the water table, factors $L$ (lateral inputs) and $C$ (water table) can be eliminated.

The values of daily rainfall were registered at the meteorological station of Corozo-Palmitas some 6 km from the studied savanna, but there are no figures for the effective precipitation; that is, what remains after subtracting the amount that is intercepted by the vegetation and subsequently evaporated without ever entering into the system. In tropical grasslands no measurements have been made of the rainfall interception, so that we must adapt the values from the rain forest. McGuiness et al. (1969) measured an interception of rainfall of 20% in a tropical semideciduous forest in eastern Panama, while Huttel (1975) indicates 10% for a comparable forest in the Ivory Coast. The amount of interception depends on the total intercepting cover and the intensity of the rainfall: the harder the rainfall, the greater the amount that penetrates the canopy. In the savanna the standing biomass is much smaller than in the tropical forests, and the precipitation is more intense. It is therefore reasonable to assume that the rainfall interception in the woody savanna is in the order of 5% of the total annual rainfall.

The variations in the soil moisture were measured by Sarmiento and Vera (1977) in 5 different soil levels. To establish the water balance we will use the data from 1974, the only complete year available. These data will tell us when there has been infiltration from one level to the next and what amount. The only variable of the model that remains unknown is transpiration.

Figure 31 shows in graphic form the variation of soil water during the year, expressed in mm and added up for two soil bands, from 0 to 125 cm and from 0 to 200 cm. The first level represents the depth explored by the herbaceous species and contains 98% of the hypogeous biomass of the system; the second level can only be exploited permanently by deep-rooted woody species, such as trees of the open savanna. Evaluation of the lower level will allow us to measure the moisture changes in each level and relate it to the transpirational behavior of one or the other group of species.

The data on rainfall for 1974 appear in table 13, followed by the calculation of the effective precipitation assuming a 5% mean interception and the total amount of water in the soil in all the horizons between the surface and 125 cm and the surface and 200 cm. Data for 31 December 1973 is also included. The respective infiltrations in the two levels were worked out by calculating the maximum amount of water that each level can retain (field capacity): it was 238 mm for the first and 218 mm for the second (these numbers

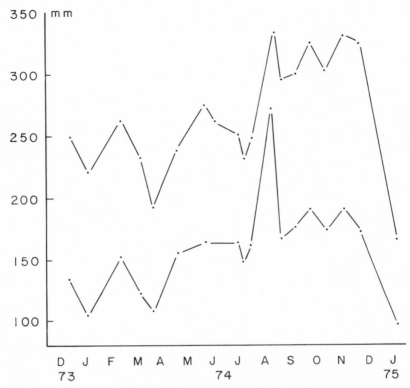

**Figure 31** Variation in the total amount of water in the soil, 1973–75, in a seasonal savanna in the llanos of Barinas (Boconoito). The bottom curve indicates the water between the surface and 125 cm; the *top* curve charts the water in the first 200 cm.

were derived from the data of field capacity and apparent density of each horizon). Finally, we calculated the losses due to transpiration by applying the formula $T = P_e - I + \Delta G_s$.

The calculations show that total transpiration reached 980 mm in 1974, if only the moisture available between the surface and 125 cm is computed; it would be 1114 mm if the water resources of the first 200 cm of the soil are used. During this period there is no evidence for infiltration below the 200 cm level, but water has descended from the upper levels to below 125 cm. On the other hand, the content of water in each of the two levels was the same at the end of the cycle as at the beginning. Thus the amount of water transpired was the same as the effective rainfall.

**Table 13.** The water balance for the year 1974 in the open seasonal savanna of *Leptocoryphium – Elyonurus* (Boconoito) in the western llanos of Venezuela (in mm).

| | 12/31 1973 | 1/1 to 1/18 | 1/19 to 2/28 | 3/1 to 3/22 | 3/23 to 4/5 | 4/6 to 5/3 | 5/4 to 6/7 | 6/8 to 6/19 | 6/20 to 7/15 | 7/16 to 7/20 | 7/21 to 7/31 | 8/1 to 8/23 | 8/24 to 9/6 | 9/7 to 9/22 | 9/23 to 10/10 | 10/11 to 10/29 | 10/30 to 11/19 | 11/20 to 12/10 | 12/11 to 12/31 | Total |
|---|---|---|---|---|---|---|---|---|---|---|---|---|---|---|---|---|---|---|---|---|
| Precipitation ($P$) | — | 0 | 80 | 15 | 26 | 65 | 145 | 29 | 105 | 8 | 43 | 117 | 73 | 84 | 112 | 79 | 126 | 3 | 0 | 1170 |
| Effective precipitation ($P_e$) | — | 0 | 76 | 14 | 25 | 62 | 138 | 28 | 100 | 8 | 41 | 168 | 70 | 80 | 106 | 75 | 120 | 3 | 0 | 1114 |
| Soil water 0–124 cm ($G_s$) | 160 | 130 | 152 | 122 | 109 | 154 | 164 | 165 | 164 | 148 | 162 | 238 | 167 | 176 | 191 | 173 | 191 | 172 | 134 | $\Delta G_s = 4$ |
| Infiltration ($I_{125}$) | — | 1 | 0 | 0 | 0 | 3 | 26 | 0 | 0 | 0 | 3 | 62 | 0 | 0 | 11 | 0 | 11 | 13 | 0 | 130 |
| Soil water to depth of 125–200 cm ($G_s$) | 117 | 116 | 111 | 111 | 82 | 85 | 111 | 97 | 87 | 83 | 86 | 148 | 129 | 124 | 135 | 129 | 140 | 153 | 112 | $\Delta G_s = -4$ |
| Infiltration ($I_{200}$) | — | 0 | 0 | 0 | 0 | 0 | 0 | 0 | 0 | 0 | 0 | 0 | 0 | 0 | 0 | 0 | 0 | 0 | 0 | 0 |
| Transpiration of plant to height of 125 cm | — | 30 | 27 | 44 | 38 | 14 | 102 | 27 | 101 | 24 | 24 | 30 | 141 | 71 | 80 | 93 | 91 | 9 | 38 | 980 |
| Transpiration of plant to height of 125–200 cm | — | 0 | 6 | 0 | 29 | 0 | 0 | 14 | 10 | 4 | 0 | 0 | 19 | 5 | 0 | 6 | 0 | 41 | | 134 |

## General conclusions

The results of the analysis of the water balance (table 13) indicate that changes in the soil water content depend on the total annual precipitation. In years with below normal rainfall the vegetation will use more water than what is incorporated into the soil, while in very rainy years a certain accumulation of water in the soil takes place resulting in a net gain for the ecosystem. In 1974, the year of our measurements, rainfall in this region was 31% below normal, or 533 mm less than average. In those circumstances transpiration, which can be equated with real evapotranspiration, was equal to effective precipitation; that is, there was neither accumulation nor infiltration.

If rainfall had been even less — if it had been an exceptionally dry year like some that happen every so often — the savanna would have continued transpiring until its water reserves became exhausted, becoming then exposed to the effects of drought. In this ecosystem the amount of water retained at 15 atmospheres in the 0–125 cm horizon is 142 mm, while in the 150 to 200 cm horizon 110 mm are retained at 15 atmospheres. There are, therefore, 252 mm between 0 and 2 m when the soil reaches the wilting point. From table 13 it can be observed that during the annual cycle both levels lose water until they are below their permanent wilting points. The upper level is in this condition from early December to early April and the lower level from the end of March to the end of July.

These facts lead to two deductions. First, during a good portion of the year there was more water than the plants used; in other words, during those months water was not a limiting factor. Secondly, in this case the physiological drought point represented by the wilting point does not have a major ecological meaning, since some species must have the ability to extract water that is under tensions greater than 15 atmospheres, or otherwise the soil moisture would not have ever reached below this level.

In circumstances of higher than normal rainfall, water will accumulate in the soil until it reaches field capacity, and then it will gradually infiltrate the lower levels as each successive horizon becomes saturated. Given the retention capacity of the various horizons between the surface and 2 m (456 mm), in order for water to infiltrate below this level effective rainfall must be higher than the amount needed to reach this figure plus an amount equivalent to the quantity of water that is transpired by the vegetation, which is about 250 to 300 mm per month. Such high values occur only in years when rainfall is much above average. For example, in Corozo-Palmitas, where the mean annual precipitation is 1703 mm, 2326 mm were recorded in 1972; in that rather wet year there were two months with rains above 250 mm and one that surpassed 300 mm. These deviations take place with certain frequency in a climate with a great annual variability in rainfall. At such times water can infiltrate the deep horizons of 2 m or more, even discounting significant losses due to lateral runoff. This additional water resource could then be used by the trees of the savanna.

According to the model that we have proposed, another possible loss of water from the system is through surface runoff, a variable that was not taken into account in our first approximation. Surface runoff takes place when rainfall is intense enough to saturate the upper soil levels, so that the rate of infiltration is lower than the rate of precipitation. Sheets of water then form on the soil surface and flow with the slope, taking with them fine soil particles in suspension. A frequent occurrence of this phenomenon produces what is known as sheet-flow erosion, or nonconcentrated erosion, since in its first phase the sheets of water do not form drainage channels. In the savanna ecosystem sheet erosion is visible from the large tussocks that often appear to be elevated a few centimeters over the level of the soil. Sheet erosion depends also on the slope, which provides the initial energy for the water to reach a certain kinetic energy, displacing and eroding the soil surface.

Flat fields with a very slight slope have little surface runoff, provided that there has not been some alteration in the normal conditions of infiltration. With maximum rainfall intensities of 20 to 30 mm/hour and slopes less than one percent, surface runoff can be calculated to be not more than 50 mm/year; normally it is probably below this figure, which is only reached in years with a high rainfall.

Direct losses due to soil evaporation would be quantitatively important only if it rained immediately after a fire, since during the rest of the year the high vegetation cover and the constant presence of a transpiring biomass prevent all but minimal losses from soil evaporation. The only direct evaporative losses are the ones due to vegetation interception, and they have already been taken into account in the concept of effective rather than actual rainfall.

A comparison of the annual and monthly values of transpiration

**Table 14.**  Precipitation, evaporation (in tank A), and transpiration data for two seasonal savannas in the llanos of Venezuela (Data for the *Trachypogon* savanna from San José and Medina, 1975).

| Savanna type and location | Annual transpiration (mm) | Annual evaporation, tank A (mm) | Annual precipitation (mm) |
|---|---|---|---|
| *Trachypogon* (Calabozo) | 1440 (burned) 1105 (not burned) | 2406 | 1839 |
| *Leptocoryphium* (Barinas) | 1115 (burned) | 2156 | 1170 |

of this open seasonal savanna in the western llanos with the results obtained by San José and Medina (1975) in a physiognomically similar savanna of the central llanos reveals quite similar figures. This is remarkable because they were obtained through entirely different methods, since San José and Medina calculated the transpiration of some of the dominant grasses in the savanna directly during selected days and extrapolated from these numbers, taking into account the leaf area index. Table 14 indicates that although our values are lower, evaporation during the period of measurements in Barinas was also lower than in the corresponding period at Calabozo. Moreover, during the year of measurement (1969) the savanna of Calabozo reached a higher aerial development than the one in Boconoito during its analyzed interval (1974).

A brief concluding comparison of the water balance of a savanna with a tropical forest will illuminate the essential similarities and differences between the two principal types of tropical ecosystems. Among the few published water balances for tropical rain forests is one by Huttel (1975) on three communities of the Ivory Coast, with a yearly precipitation of 1800 to 1950 mm. The annual transpiration found in these three communities was in the range of 965 to 1000 mm, which represents between 51 and 54% of the precipitation. On the other hand, the total infiltration to horizons lower than the one tested (232 cm) was between 27 and 36% of the rainfall, that is, quantities greater than 600 mm. We have already pointed out that effective rainfall was equivalent to 90% of rainfall above the canopy.

Annual values of transpiration in the savanna are very similar, despite the enormous structural differences between these ecosystems. The much greater foliar surface of the rain forest appears not to produce greater transpiration. The most significant difference in the two ecosystems is found in the high values of infiltration in the rain forest, indicating an excess of rainfall in the forest with respect to the savanna. This amount of infiltration can only be reached in the savanna in years with above average rainfall, that is, in years with a rainfall that is equivalent to that of the rain forest.

# 6

# Nutrient economy

## The savanna as an oligotrophic ecosystem

One of the explanations given for the stability of the savanna
ecosystems in the American tropics is that they are found on soils
that are extremely poor in nutrients and therefore cannot sustain
any type of woody formation. This view of the savanna as an area of
oligotrophic vegetation was put forth by several scientists working
in Surinam (Pulle, 1906; Lanjouw, 1936), especially when these
ecosystems are found on podzolic soils developed over so called
"white sands" (Heyligers, 1963). In Brazil too the various kinds of
cerrados have been associated with the least fertile soils (Waibel,
1948; Alvim and Araujo, 1952; Arens, 1958).

Arens (1963) maintains that the scleromorphism of the trees and
shrubs of the cerrado is produced by the insufficiency of calcium,
phosphorus, sulphur, and nitrogen in the soil, as well as of the
oligoelements zinc and molybdenum. He compares this morphoecological response with that of the serpentine species, growing on
soils that are also clearly oligotrophic.

In 1963 Ranzani summarized the then-available information on
the fertility of cerrado soils and concluded that in most cases the
cation exchange capacity (C.E.C.) is less than 4 meq/100 g, with
saturation rates of less than 50% and frequently less than 25%. That
is, cerrado soils are very poor in nutrients.

It is generally recognized that throughout the humid tropics soils
are very poor in nutrients, especially in the case of zonal soils that
have been subjected to prolonged periods of leaching and ferralization (Buringh, 1970). In special cases such as senile profiles with

prolonged ferralitic evolution or tropical podzolic soils, this poverty of nutrients is still more evident. The same generalization applies to profiles formed over sedimentary material that had been previously impoverished in pedogenesis (recovered alluvial materials), which is often the case in the llanos.

It has also been maintained that the recurrent fires in the savannas contribute to the impoverishing of the soils, by causing the loss of volatile elements during the fire and washing away of the ashes by wind or surface flow. Whatever the reasons for the low nutrient value of the soils, the real issue is what adaptive mechanisms allow the herbaceous and woody species of the savanna to maintain a nutrient balance in this highly oligotrophic environment.

This chapter will investigate the economy of some mineral elements in the savannas in order to discover the general role of the nutrients in these systems and also their role in determining the morphoecological and physiological characteristics of the plant species, such as the frequent scleromorphy of the woody species. The internal recycling of minerals will also be a subject of investigation. The morphogenetic processes that have produced the chemical characteristics of the savanna soils will be compared to tropical soils that have a different type of vegetation. An analysis of the nutrient content of the vegetation and the principal species will be followed by a discussion of the cycle of some nutrients in the ecosystem and of special problems related to other chemical elements.

# The genesis of the relief and its relation to the characteristics of the landscapes and soils

Both the neotropical and the African savannas in the Guinea-Zambesi area are found exclusively in humid tropical climates (*Af, Am,* and *Aw*), and occupy soils with certain common characteristics derived from high rainfall and temperature. There are three other factors that may have played an important differentiating role in pedogenesis: topography, parental material, and age. Within a given climatic region, these three factors, together with plant cover, control the evolution of relief; there is thus a very close correspondence between geomorphogenic and soil formation processes, and the characteristics of natural ecosystems.

Most of the American savannas occupy plains of relatively flat or slightly undulating topography that has diverse origins. The principal influence that relief has over the ecosystems is on the regulation of the drainage conditions, and ultimately on the water balance. This influence in turn translates into important consequences regarding the chemical and nutritional characteristics of the soils. The factors that produce the relief, through their action on pedogenesis, have indirectly determined the physico-chemical characteristics of the soils. Summaries of five relief studies follow.

The first example is the Quaternary sedimentary plains, where the savanna ecosystems developed over different geomorphological units that can be differentiated by their chronology and by manner of their formation. The original materials, which came from erosional zones that fed the sedimentary basins, suffered after their deposition the impact of soil formation processes typical of this morphoclimatic zone, both in terms of elapsed time and of the drainage conditions determined by the form of the relief and the geomorphological position of each habitat within it.

A very illustrative analysis of the interrelationships of geomorphological units, forms of relief, topographical position, and pedogenesis, was performed by Zinck and Urriola (1968) working in the cut-and-fill terraces of the Rio Guarapiche in the state of Monagas (Venezuela). In this valley there are four levels of alluvial accumulations that show very clearly the precise correspondence between the age of the deposit and the relative degree of evolution of the profile. The oldest terraces, corresponding to the early Quaternary (Q3 and Q4), found in a high topographic position (bank) with good drainage conditions, have the most lixiviated soils with the most developed and deepest argillic horizon and the most accentuated desaturation, having reached in their evolution the state of ultisols, or even oxisols (table 15 and figure 32). In contrast, in the subactual terrace (Q1), the young and well-drained soils (inceptisols) support a tropical rain forest, in contrast to the other three levels where only savannas are found. The Q1 soil shows a much more favorable nutrient balance, owing to the incipient weathering and pedogenesis as well as the vegetable cover, which tends to reinforce the mesotrophic character of the soils. The Q2 terrace, with alfisols in the bank position, is intermediate in terms of the degree of evolu-

**Table 15.** Some soil characteristics of five well-drained terraces on the Guarapiche River (State of Monagas) at a depth of 50 cm (data from Zinck, 1970).

| Relative age | Soil series and order | Clay (%) | Base saturation (%) | pH | C.E.C. (meq/100 g clay) |
|---|---|---|---|---|---|
| Q0 | Guatatal (entisol) | 12 | 100 | 7.8 | 130 |
| Q1 | San Feliz (inceptisol) | 18 | 90 | 6.3 | 90 |
| Q2 | El Tomate (alfisol) | 9 | 22 | 6.2 | 35 |
| Q3 | El Zamuro (ultisol) | 18 | 38 | 4.3 | 65 |
| Q4 | Sabaneta (ultisol) | 22 | 20 | 4.7 | 25 |

tion of the profile and in the desaturation and impoverishment of nutrients.

This progressive evolution of soils in deposits of increasing age is not necessarily a sustained unidirectional process that has been taking place since the deposition of soils to this day. On the contrary, all the profiles show signs of a polycyclical evolution. This development has been caused either directly by the large climatic oscillations that characterized the Pleistocene of North and South America, or indirectly by the changes in drainage conditions, consequence of the successive variations in the base levels (marine regressions and transgressions that accompanied the glacial and interglacial periods). Moreover, in many of these alluvial zones the geomorphogenetic fluvial cycles were interrupted by arid cycles, during which the superficial deposits were remodelled into new forms and characteristics that modified the preceding pedogenetic evolution.

In these chronological sequences associated with the successive depositions in alluvial regions such as the llanos, the more evolved soils present a very deficient nutrient status. Table 16 presents soil

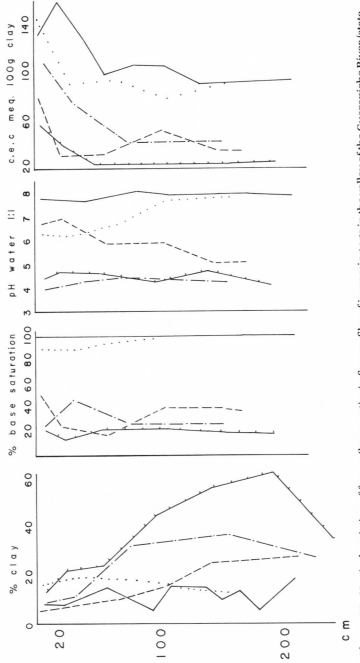

**Figure 32** Vertical variation of four soil properties in five profiles of increasing age in the valley of the Guarapiche River (state of Monagas). *Thick continuous line:* Guatatal series (Q0); *dotted line:* San Felix series (Q1); *dashed line:* El Tomate series (Q2); *dot-and-dash line:* El Zamuro series (Q3); *fine broken line:* Sabaneta series (Q4). For more information on these series see table 15. (Data from Zinck, 1970.)

**Table 16.** Mean and extreme values of several soil characteristics in ten savanna soil profiles on oxisols on the Q4 terrace in the Barinas piedmont.

| Horizon (cm) | pH | m.o. (%) | Clay (%) | P (p.p.m.) | C.E.C. (meq/100 g) | Ca (μeq/100g) | Mg (μeq/100 g) | K (μeq/100 g) |
|---|---|---|---|---|---|---|---|---|
| 0–10 | 4.90 | 1.98 | 27 | 2.20 | 3.49 | 63 | 465 | 618 |
|  | 4.75–5.10 | 2.36–1.52 | 34–20 | 4.20–0.70 | 4.50–2.50 | 249–0.0 | 1168–272 | 1112–384 |
| 10–50 | 5.13 | 1.33 | 32 | 1.55 | 2.98 | 10 | 235 | 380 |
|  | 4.95–5.40 | 2.05–0.96 | 47–24 | 2.38–0.70 | 3.75–2.50 | 49–0.0 | 319–164 | 818–179 |
| 50–80 | 5.07 | .86 | 34 | 1.27 | 2.87 | 5 | 183 | 266 |
|  | 4.60–5.30 | 1.17–0.0 | 45–25 | 2.38–0.70 | 3.50–2.12 | 49–0.0 | 310–92 | 818–77 |

nutrient data that correspond to the average of ten profiles on the Q4 accumulation in the llanos de Barinas. The soils of the seasonal savannas that are found in this zone are consequently extremely poor in almost all macronutrients.

A second form of relief, composed of older alluvial deposits from the end of the Pliocene and beginning of the Pleistocene, remained elevated over valleys, escaping in this way from the influence of later alluvial cycles. The relief of these mesas or high plains, with altitudes in the order of 150 to 350 m, was progressively modelled so as to form various piedmont steps. These formations frequently exhibit several levels of lateritic duricrusts formed under different conditions, which later acted as forms of resistance to erosion and thereby directed the evolution of the relief.

The mesa formation underlies a great portion of the central and eastern llanos of Venezuela and includes a strip in the northern border of the Guyana massif. The geomorphology, soils, and vegetation of these mesas were analyzed in several works (Sarmiento and Monasterio, 1969, 1971; Blanck et al., 1972; Blanck, 1976). Another high plain runs from the south of the state of Apure in Venezuela through the llanos of Colombia and enters the Amazonas rain forest to the south of the Rio Guaviare (FAO report, 1964; Comerma and Luque, 1971). Likewise, some of the savannas of the Rio Branco (Roraima Territory) in Brazil and the savannas of Rupununi in the south of Guyana occupy accumulations of this age and type of relief (Sinha, 1968).

The soils that evolved over in situ materials in the Plio-Pleistocene mesas and high plains, under conditions of good drainage, have reached the level of ultisols, or more generally of oxisols; while those that were formed over internal redeposits in the glacis, although less evolved pedogenetically, are equally poor in nutrients owing to prior impoverishment of the recovered material that formed the substrate of the new profile.

A third formation is the elevated massif of complex geological structure that forms the greater part of the central planalto of Brazil, characterized by Ab'Saber (1963, 1971) as the domain of the *chapados* landscape with a cerrado vegetation. Here a deep saprolite covers what are usually Precambric or Paleozoic rocks. Here too, and with greater reason, the soil profile has acquired senile characteristics, with a predominance of very impoverished oxisols, as we will see further on.

A special situation may be found on these continental mesas when the geological substrate is a more or less friable sandstone, whose weathering under high temperature conditions, high moisture, and good drainage led to the development of a podzolic profile. This occurs, for example, on the sandstone of the Roraima formation (Precambrian). Provided that the mesas *(tepuyes)* do not surpass 1500 m, their completely lixiviated podzols support a special type of ecosystem, dominated by very scleromorphic grasses and monocots ("apparent savannas"; see Steyermark et al., 1976).

A fourth case, similar in many respects to the preceding one, is encountered when the soil has been formed from unconsolidated materials originating in the erosion of sandstone massifs. This is the origin of the extensive "white sands" that border the Guyana massif both to the south in the region of the Negro and Amazonas rivers and on the coast from Guyana to Amapa, Brazil. Over this unconsolidated material tropical podzols have also developed, sustaining a very poor type of rain forest, the Amazonian *caatingas* (Klinge, 1967; Klinge and Herrera, 1977) or savanna (Egler, 1960; Heyligers, 1963).

The fifth and less frequent case is that of the savannas that cover a landscape of low, rolling hills, as on the continental slope of the Cordillera de la Costa in Venezuela, or on the northern border of the Guyana shield to the south of the Orinoco River. Under these topographic conditions, in contrast to the other four, the soils rejuvenate gradually by erosion and superficial removal, so that their chemical characteristics depend in large measure on the type of substrate, which is in general poorer over granitic rocks and sandstones, and relatively richer in mineral elements over metamorphic, basic igneous, or limestone rocks.

In summary, the poorest soils where savannas grow are found in the alluvial plains of Early Pleistocene (Q3 and Q4); Plio-Pleistocene high plains; very old erosion surfaces and pediplains; "white sands." The best tropical soils are found over young volcanic materials; metamorphic, basic igneous, and limestone rocks; recent alluvial deposits; substrate with deficient drainage.

In this last case (hyperseasonal savannas and marshes) it is physical factors that act as the fundamental limiting factors for the growth of plants, since the lack of drainage has impeded lixiviation and the consequent nutrient impoverishment has affected only the uppermost part of the soil profile.

# Chemistry of savanna and other tropical soils

Greenland and Kowal (1960) were the first to describe the nutrient content of the vegetation and the soil in a tropical forest in Ghana ("moist semideciduous forest"). In latosols, the horizons between 0 and 5 cm, and between 5 and 30 cm contain, respectively, the following nutrients: total nitrogen, 0.198 and 0.101%; assimilable phosphorus, 8.4 and 2.0 p.p.m.; exchangeable potassium, 0.46 and 0.33 meq/100 g; exchangeable calcium, 5.52 and 2.32 meq/100 g; exchangeable magnesium, 1.11 and 0.59 meq/100 g; cation exchange capacity, 9.22 and 6.10 meq/100 g. This describes a relatively poor soil, with particularly low quantities of phosphorus and potassium.

Stark (1970) analyzed the nutrient content in the soils and related it to the various types of vegetation growing over white sands in Surinam, from a mature forest (30-m high *Dimorphandra* forest) to a savanna forest (lower and more open), a savanna woodland of 10–12 m in height, a savanna shrubland with a few species of trees 3 to 5 m high, and finally to areas of white sands almost completely devoid of vegetation. In addition she compares the data from the white sands (an ever-wet area with 2000 to 2400 mm of rainfall) with data obtained in latosols that maintain a tropical forest in Surinam, and with tropical podzols developed over white sands in the Brazilian Amazonia (Stark, 1971a,b). The latosols and the species that prosper on them surpass the podzolic soils and the species growing on them in the total amount of nine nutrients. The latosols under the rain forest have in their uppermost horizon three to six times more potassium, calcium, and magnesium than the podzolic soils under the forest, whether in Brazil or in Surinam; while the amount of these elements in the same superficial horizon is much lower in the white sands that are almost devoid of vegetation. A gradient of nutrient impoverishment can be detected in the sequence of formations on the white sands, from the forest to the open formations. It is not clear whether the vegetation gradient is a consequence of the dystrophism resulting from the accelerated washing away of the soils after the disappearance of the original forest, or whether it is a response to the original nutrient poverty of the soils.

Stark's study of the nutrients of the soils and the vegetation of the Brazilian Amazonia was performed in soils under different types of forest, including mature forests *(matas da terra firme)*, flooded rain forests *(igapos)*, and early successional stages *(capoeiras)* in the area of Manaus, Madeira, and the confluence of the Negro and Branco rivers. Her data reveal a notably low fertility. Between 0 and 62 cm the soil pH ranged from 3.6 to 4.6; the cation exchange capacity was between 7 and 26.3 meq/100 g; the base saturation varied from 2.7 to 28.5%, while the quantities of exchangeable calcium and potassium were 5 to 6 $\mu$eq/100 g and 12 to 629 $\mu$eq/100 g respectively. The low nutrient content of the Amazonian forest contrasts so starkly with the development of the vegetation that the author proposes the hypothesis of a direct cycling of nutrients, which would pass from the decomposing litter through mycorrhizal fungi to the tree roots, bypassing the mineral soil (Went and Stark, 1968).

Anderson (1981) arrives at similar conclusions when discussing the possible origins of the evergreen sclerophyllous vegetation *(caatinga)* on white sands in Brazilian Amazonia. Comparing the soil properties of the nonhydromorphic white sand entisols near Manaus, on which the Amazonian caatinga is found, with fine-textured yellow oxisols that maintain a richer rain forest, he shows that the topsoil of the entisol has one-third of the carbon, one-eighth the nitrogen, one-fourth the cation exchange capacity, three-fifths the potassium, half the phosphorus, one-fifth the calcium, and one-tenth the magnesium of the oxisol, which already is a very poor soil.

Coutinho and Lamberti (1971) analyzed three profiles in the region of the Rio Negro in the Brazilian Amazonia: a mature forest *(mata da terra firme)*, a recently cultivated plot in a deforested parcel in the same forest, and a flooded rain forest *(igapo)*. All three soils turned out to be very acid; in the first two types the pH was below 4. Likewise, the base saturation rate was extremely low throughout the three profiles (2 to 10%), and so were the values of the exchangeable cations: $Ca^{++}$ (0.05 to 0.6 meq/100 g); $K^+$ (0.02 to 0.32 meq/100 g); and $Mg^{++}$ (0.55 to 0.05 meq/100g). According to the authors these soils are representative of a large area of the Amazonian forest, implying a very low nutrient status in the soils of the most extensive area of rain forest on earth.

Williams and his colleagues (1972) also studied similar soil profiles. The mature forest over well-drained soil *(mata da terra firme)* prospered on a yellow caolinitic latosol, developed over a river terrace of the Early Pleistocene. The profile had a pH of 4.0 or less; the cation exchange capacity varied between 8 meq/100 g on the surface to less than 3 meq/100 g at 75 cm; exchangeable potassium varied between 0.05 and 0.03 meq/100 g, and no exchangeable magnesium or calcium was found. Soluble phosphorus was also extremely low (0.36 ppm). In the manioc field, burning produced a slight increase in calcium and magnesium, but the soil continued being exceptionally poor. The soil of the floodable forest *(igapo)* was humic glei, poorly drained, with a pH also below 4.0, while the values of the cation exchange capacity and of the exchangeable cations were as low as those of the mature forest.

Golley et al. (1975) elaborated one of the most detailed studies on nutrient cycles done to date in low-altitude tropical American forests. They found in the soils of the semideciduous forest of Darien (Panama) a much higher nutrient content than what is known for other tropical forest formations. Thus these authors indicate a cation exchange capacity in the 0–30 cm horizon of 50 meq/100 g, with a base saturation of 86%. More than 90% of the exchangeable cations are calcium and magnesium. The causes for the discrepancy are geological and topographical, as the parent material is a Miocene shale ("savanna beds") intercalated with dolomite and calcareous sandstone. The residual soil developed over this material forms a deep saprolite (8 to 14 m) and is a black clay soil formed primarily by montmorillonite, a clay with a high cation exchange capacity. In addition, the high clay content results in slow drainage, which impedes the leaching of the soil. In addition, as Golley and his colleagues point out, although the rainfall is high (2000 mm/year), there is a three-month dry period with less than 50 mm/month of rainfall, which diminishes the losses of mineral nutrients by deep infiltration. In conclusion then, although the great majority of the soils developed under tropical rain forests are markedly oligotrophic, there are exceptions caused by specific conditions of climate, relief, and geological substrate.

Research on savanna soils found them even poorer in nutrients than the tropical forests. Alexander (1973) compared the soils of forests and savannas in northeastern Nicaragua (with about 3000 mm of annual rainfall), both lying on Early Pleistocene alluvial deposits in the plains of the Caribbean piedmont. The chemical analysis of a soil series from the seasonal forest and the seasonal savanna, both ultisols, shows that the nutrient content of these forest soils is higher than that of the Amazonian forest and much higher than that of the savanna soils (table 17). In the first 15 cm, the forest soils have five times more nitrogen than the savanna soils,

**Table 17.** Comparison of the nutrients in a forest and a savanna soil in Nicaragua, both profiles on the same piedmont deposit (data from Alexander, 1973).

| Horizon | pH | Organic C (%) | Total N (%) | C.E.C. (meq/100 g) | Base saturation (%) | Ca | Mg (meq/100 g) | K |
|---|---|---|---|---|---|---|---|---|
| *Ultisol-Rain forest* | | | | | | | | |
| A11 | 5.5 | 5.3 | 0.53 | 36.7 | 23 | 4.29 | 3.53 | 0.68 |
| A12 | 5.4 | 4.1 | 0.39 | 26.2 | 17 | 1.90 | 1.91 | 0.43 |
| A3 | 5.2 | 2.7 | 0.26 | 19.1 | 6 | 0.26 | 0.79 | 0.15 |
| B1 | 5.2 | 0.9 | 0.12 | 11.0 | 7 | trace | 0.56 | 0.16 |
| B21 | 5.3 | 0.5 | 0.08 | 14.0 | 9 | 0.05 | 1.07 | 0.07 |
| B22 | 5.3 | 0.2 | 0.07 | 12.3 | 9 | 0.11 | 0.89 | 0.05 |
| B23 | 5.2 | 0.2 | 0.06 | 17.9 | 7 | 0.11 | 1.06 | 0.05 |
| *Ultisol-Savanna* | | | | | | | | |
| A1 | 4.6 | 2.1 | 0.14 | 10.3 | 4 | trace | 0.29 | 0.05 |
| A2 | 4.8 | 1.0 | 0.07 | 6.2 | 4 | trace | 0.23 | 0.04 |
| B21 | 5.1 | 0.9 | 0.08 | 8.1 | 4 | trace | 0.25 | 0.04 |
| B22 | 5.1 | 0.5 | 0.05 | 9.3 | 3 | trace | 0.29 | 0.03 |
| B23 | 5.3 | 0.3 | 0.02 | 8.4 | 3 | trace | 0.25 | 0.03 |
| B24 | 4.9 | 0.1 | 0.01 | 8.1 | 4 | trace | 0.29 | 0.04 |

sixty times more calcium, ten times more magnesium and potassium, five times the base saturation, and four times the cation exchange capacity. The only difference between the two profiles, the vegetation they maintain, must have taken several thousand years to evolve. Hence the savannas in this region are probably formations that antecede the arrival of man in Central America.

Markham and Babbedge (1979) found a similar situation in Africa. Soil samples were analyzed for major plant nutrients along soil catenas on the forest-savanna boundary in west-central Ghana. Vegetation types were found on distinct catenary positions associated with characteristic soil types. Levels of total phosphorus, exchangeable calcium and potassium, water content, and pH in the topsoil were all found to be significantly higher under the forest than under the savanna.

In his latest survey of the broad characteristics of the soils of the Brazilian cerrado, Ranzani (1971) presented statistics on the range and distribution of some edaphic variables in 112 horizons that correspond to 23 different complete profiles of soils under different types of cerrado. Figure 33 shows his data for pH, C.E.C., and sum of exchangeable bases (S). The pH value lies between 4.0 and 6.0 in 91% of the samples, and between 4.0 and 5.0 in 51% of them. The C.E.C. is between 2 and 6 meq/100 g in 89% of the cases, while S is below 1.8 meq/100 g in 89% of the cases. This sampling, which included the principal types of soils under the Brazilian cerrado, confirms their nutrient poverty.

Askew et al. (1971) studied various soil profiles under cerrado and dry evergreen forest in the Serra do Roncador, Mato Grosso, Brazil. The figures for the soils of the cerrado are very low. For instance, exchangeable calcium reached only 11.78 $\pm$ 1.38 kg/ha. This low figure is nevertheless higher than the respective value for the dry forest, which may thus be classified as another tropical ecosystem that maintains itself over what the authors call "dystrophic soils."

In his revision of the vegetation of the cerrado Eiten (1972) synthesized a great deal of the then available information on the chemical characteristics of the soils, confirming the data of Ranzani about pH, cation exchange capacity, and base saturation. He also calculated that in superficial horizons (to 10–20 cm of depth) the nitrogen content is between 0.01 and 0.2%; that of calcium from 0.2 to occasionally 1 meq/100 g, and the exchangeable potassium reached an average of 0.1 meq/100 g.

Goodland and Pollard (1973) studied the savanna formation of the

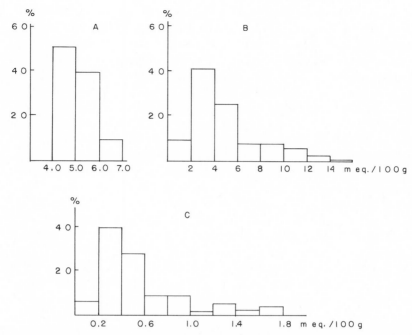

**Figure 33** Frequency distribution of pH *(A)*, cation exchange capacity *(B)*, and total base exchange *(C)* in 112 soil horizons under cerrado vegetation in Brazil. (From Ranzani, 1971.)

Minas Gerais triangle in the center of Brazil and compared the nutrient content in 110 samples of soil, arranged according to four fundamental physiognomic types: grasslands with dispersed, low shrubs *(campo sujo)*, closed savannas *(campo cerrado)*, the cerrado proper or woody savanna, and the scleromorphic forest or cerradao. In the soil level between 0 and 10 cm, the exchangeable potassium increases in this vegetation gradient from 0.08 to 0.17 meq/100 g of soil; the nitrogen from 0.07 to 0.1% and phosphorus from 0.024 meq/100 ml to 0.067 meq/100 ml. This significant correlation between potassium, phosphorus, and nitrogen, and woody plant density in the cerrado cannot however be interpreted as the direct cause and effect. First it must be established whether the nutrient difference itself is cause or effect of the variability in the tree cover in the savanna.

In any case, and in confirmation of these data, Goodland and Ferri (1979) also found that the soils under cerrado vegetation were sharply dystrophic, with quite low levels of calcium, potassium, and phosphorus. Galrao and Lopes (1980) conducted a still wider sampling (518 topsoils from throughout the cerrado area) and reported that 96% of the samples were below the critical level of exchangeable calcium for crops, and that 90% were below the critical level for magnesium, 99% below that for phosphorus, and more than 90% were above the critical level of tolerance for aluminum.

As for the hyperseasonal savannas, Ahmad and Jones (1969) studied the content of soil nutrients and of different plant species in the savanna of Aripo in Trinidad.The soil is an ultisol (plinthaqualt) that is periodically flooded during the rainy season (rainfall greater than 2500 mm), and the open savanna is dominated by *Byrsonima crassifolia* and different species of *Panicum* and *Paspalum*. The fertility of this soil is very low, with values in the superficial horizon of 0.048% of nitrogen, 0.4 to 0.5 meq/100 g of calcium, 0.2 to 0.8 meq/100 g of magnesium, 0.02 to 0.12 meq/100 g of potassium, and 1 to 2 p.p.m. of extractable phosphorus. This is a soil that is notably poor in nutrients, in spite of the deficient drainage conditions.

Another analysis of the nutritive characteristics of soils under savanna and forest was elaborated by us on the basis of our own data as well as those in the literature in different areas of the llanos of Venezuela and Colombia. We examined 182 samples of savanna soils, using only the uppermost described horizon of the respective profiles, since it is the one that shows the greatest differences in function of the vegetation. Of these 182 analyses, 102 correspond to soil series in the llanos of Colombia (one profile in each of the series of the FAO inventory, 1966), 70 are from the western llanos of Venezuela (Silva and Sarmiento, 1976 b), and 10 were taken from the soil study of the Santo Domingo-Paguey zone in the same region of the Venezuelan llanos (Zinck and Stagno, 1965). The forest soils included 85 profiles: 15 from the llanos of Colombia (FAO, 1965), 38 from the Ticoporo I study in the western llanos of Venezuela (Blanck et al., 1970), and 32 from the area of Ticoporo II, adjacent to the first one (Castillo et al., 1972). Table 18 shows that the forest soils of the llanos, in comparison to savanna soils, have 33% more nitrogen; almost double the base saturation; 80% more exchangeable potassium; five times more exchangeable calcium, and 13 to 15% more cation exchange capacity and extractable phosphorus.

**Table 18.** Comparison of some chemical characteristics of rain forest and savanna soils in the llanos of Colombia and Venezuela.

| Location | Horizon (cm) | Total N (%) | C.E.C. (meq/100g) | Base saturation (%) | K (meq/100 g) | Ca | Total P (p.p.m.) |
|---|---|---|---|---|---|---|---|
| | | | *Savanna* | | | | |
| Colombia (102 profiles)[a] | 0–12 | 0.208 | 13.0 | 17.2 | 0.22 | 0.81 | 6.66 |
| Barinas (70 profiles)[b] | 0–10 | 0.051 | 4.6 | 20.5 | 0.08 | 0.67 | 1.29 |
| Barinas (10 profiles)[c] | 0–13 | 0.080 | 6.1 | 58.7 | 0.14 | 2.14 | — |
| Average (182 profiles) | 0–11 | 0.140 | 9.3 | 20.5 | 0.16 | 0.83 | 4.40 |
| | | | *Rain forest* | | | | |
| Colombia (15 profiles)[a] | 0–15 | 0.203 | 12.3 | 32.4 | 0.26 | 1.87 | 5.22 |
| Ticoporo I (38 profiles)[d] | 0–19 | 0.184 | 11.5 | 55.6 | 0.33 | 4.85 | 5.18 |
| Ticoporo II (32 profiles)[e] | 0–14 | 0.171 | 8.4 | — | 0.27 | 4.27 | 5.02 |
| Average (85 profiles) | 0–17 | 0.184 | 10.5 | 49.0 | 0.29 | 4.1 | 5.10 |

*Note on data:* (a) FAO, 1965; (b) Silva and Sarmiento, 1976; (c) Zinck and Stagno, 1965; (d) Blanck et al., 1970; (e) Castillo et al., 1972.

# The principal nutrients in the vegetation

In one of the first studies on the subject of nutrient content in tropical vegetation, Bartholomew et al. (1953) looked at different successional stages of a tropical forest in Yangambi, Zaire, and determined the biomass and various nutrients contained in it. The communities were three grasslands dominated respectively by *Panicum maximum*, *Setaria sphacelata*, and *Cynodon dactylon*, and two forest renewals of five and eighteen years of age. The average data of the herbaceous stages and of both woody communities are depicted in table 19. The data show that the percentages of nutrients keep augmenting significantly along the succession, so that by the eighteen-year stage, characterized by the dominance of *Musanga cecropioides* (gap species of fast growth), the values are high, especially those of nitrogen and potassium.

Greenland and Kowal (1960) determined the content of five elements — nitrogen, phosphorus, potassium, calcium, and magnesium — in a fifty-year-old secondary, semideciduous forest in Ghana. The distribution of the biomass and the nutrient content of the various vegetation parts were analyzed. Table 19 reproduces the data of Greenland and Kowal for the part of the vegetation that is richest in nutrients — leaves and branches less than 5 cm long. The amount of analyzed elements seems relatively high, both in comparison with the soil content of these elements and with the amounts of these elements found in temperate vegetation. Indeed, the authors point out that the percentages of nutrients in the tropical forests are higher than in temperate forests of the same age, whether broad-leaved or coniferous, and they are even comparable to the values that these communities reach at 100 years of age.

Gottlieb and others (1966) analyzed the mineral composition of leaves of trees of forests and cerrados in Brazil, comparing in each case species of the same genus that grow in one or the other type of formation. The authors concluded that there are no differences in the accumulation of boron, calcium, potassium and zinc between the equivalent pairs of species, despite important nutrient differences between the soils. The analysis of 21 pairs of species provided no evidence that the leaves of the forest trees accumulate more nutrients than the trees of the cerrados, with the exception of sodium, which is systematically lower in species of the cerrado.

Working in Surinam and in the Brazilian and Peruvian Amazonia, Stark (1970, 1971b) analyzed the composition of the leaves,

**Table 19.** Nutrients in trees and tropical forests.

| Ecosystems and Locations | N (%) | P (%) | K (%) | Ca (%) | Mg (%) | Observations | Sources |
|---|---|---|---|---|---|---|---|
| 3-year-old secondary rain forest succession Zaire[a] | 0.96 | 0.10 | 0.82 | 0.44 | combined | Mean aerial biomass of 3 communities | Bartholomew et al. (1953) |
| 18-year-old secondary rain forest succession Zaire | 2.20 | 0.12 | 1.24 | 1.18 | combined | Leaves | Bartholomew et al. (1953) |
| 50-year-old secondary forest on latosol Ghana | 1.88 | 0.13 | 0.76 | 1.90 | 0.26 | Leaves and branches, 5 cm diameter | Greenland and Kowal (1960) |
| Rain forest on latosol Brazil | 1.52 | 0.04 | 0.17 | 0.30 | 0.18 | Leaves on soil, mean of 51 samples | Klinge (1977) |
| Rain forest on latosol Surinam | 0.62 | 0.12 | 1.42 | 0.50 | 0.32 | Leaves, mean of 20 species | Stark (1970) |
| Rain forest on podzol Brazil | 2.83 | 0.25 | 0.65 | 0.25 | 0.22 | Leaves, mean of 20 species | Stark (1970) |
| Rain forest on podzol Surinam | 2.04 | 0.14 | 0.15 | 0.80 | 0.24 | Leaves, mean of 20 species | Stark (1970) |
| Open shrubland on podzol Surinam[a] | 0.50 | 0.06 | 0.15 | 0.68 | 0.38 | Leaves, mean of 20 species | Stark (1970) |
| 100-year-old semideciduous forest on black tropical clay Panama | 1.40 | 0.12 | 1.14 | 1.66 | 0.19 | Total biomass, rainy season | Golley et al. (1975) |
| *Mangifera indica*, (mango) cultivated[a] | 1.11 | 0.09 | 0.95 | 1.23 | 0.33 | Leaves | Bazilevic and Rodin (1966) |
| Mean of trees and forests | 1.78 | 0.13 | 0.79 | 0.90 | 0.23 | | |

a. Not integrated into the mean.

wood, bark, and roots in numerous tree species in different types of tropical forests, on podzols and latosols (see table 18). Of all the plant organs examined, the leaves contained the greatest amount of mineral elements. For nitrogen, phosphorus, and magnesium, the order of magnitude of the values was similar to that of the African forests previously mentioned, while the percentages of potassium and calcium were lower. On the other hand, when the leaves of these tropical trees were compared to the leaves of broad-leaved temperate trees, it was found that the amount of iron, manganese, potassium, calcium, and magnesium was lower, and that of sodium much lower in the tropical trees, while the content of phosphorus was equivalent to that of temperate forest trees.

In the semideciduous forests of Darien (Panama) growing over relatively rich soils, Golley et al. (1975) found the following total mineral contents for an average of different organs and plant parts (in ppm): 14,000 for nitrogen, 16,600 for calcium, 11,400 for potassium, 1200 for phosphorus, 1900 for magnesium, 1040 for aluminum, 200 for sodium, and 133 for iron. The authors point out in their conclusions that: active growing tissues such as leaves, flowers, and fruits have a higher nutrient content than less active tissues such as wood; understory species have a greater mineral concentration than canopy ones; secondary formations in active development have a higher concentration of mineral nutrients than mature forests. On the other hand, the semideciduous moist tropical forest is richer in nutrients than the moist tropical montane forest found on mountain slopes above 250 m. The combined data from all types of forests reveal that of the analyzed mineral elements (excluding nitrogen, which was not included in the analysis), calcium and potassium turned out to be the most abundant, followed by magnesium, phosphorus, sodium, and aluminum. The authors also indicate that the order of abundance in the vegetation is not the same as that in the soils, in which calcium and magnesium are the most abundant mineral nutrients.

Klinge (1977) determined the nutrients in a seasonal tropical forest near Manaus (Brazil), in the central Amazonian region. The proportion of nutrients in leaves is reproduced in table 18, with potassium and calcium in relatively low concentrations.

Finally, we have added in table 18, for comparative purposes, the values given by Bazilievic and Rodin (1966) for nutrients in the leaves of the mango (*Mangifera indica*), as an example of a tropical cultivated tree that is perennial and sclerophyllous. The values of nitrogen and phosphorus are low, both in comparison with other cultivated species and in reference to the trees of tropical forests.

After this review of the mineral elements in the trees of the tropical forest, we will now turn to the nutrient content of typical savanna trees. Montes and Medina (1977) analyzed the nutrient content of trees from the savanna of *Trachypogon* in Calabozo, Venezuela, comparing especially perennial with deciduous trees. Their data (see table 19) indicate that the perennials accumulate significantly lower amounts of nitrogen and phosphorus, and slightly higher quantities of potassium, calcium, and magnesium. It should be pointed out that these perennial trees are exclusively savanna species, whereas the three deciduous species are rather characteristic of the deciduous forest and are seldom found in woody savannas, so that it is safe to generalize that the typical savanna trees have a lower content of nitrogen and phosphorus than the more widely growing species found both in the savanna and in other formations.

The mineral content of the leaves of the tall trees of the woody savannas known as *miombos*, which cover great areas of the savanna belt from Zimbabwe and Angola to Uganda and Kenya, was analyzed by Ernst (1975). He found important differences in the accumulation of macro- and micronutrients in 17 tree species that grow in two miombo communities in Zimbabwe. For example, the total average ash content was of 5.07%, while the extreme values were 1.89 and 10.14%. The interspecific variation of calcium is equally high (mean, 376 g/atom/g dry weight; extremes 35 and 815 g/atoms/g dry weight); for potassium (mean, 116 g/atom/g dry weight; extremes, 29 and 346 g/atom/g dry weight) and for phosphorus (mean, 42 g/atom/g dry weight; extremes, 13 and 90 g/atom/g dry weight). He concludes that even when these trees are growing on the same greatly impoverished soil (oxisol), each species of the miombo has its own capacity to accumulate different nutrients in the leaf biomass.

Comparing the mineral content of trees in this community with tree species of the deciduous communities of the northern hemisphere, Ernst points out that the range of values is the same in both groups, except for nitrogen, which is systematically lower in the tropical ecosystem. Even in species of Leguminosae that have radical nodules the nitrogen content is very low, which suggests that perhaps there are no favorable conditions for nitrogen fixation in this woody savanna.

Finally, Ernst follows the seasonal variations of the leaf nutrients in the three dominant species of the miombo: *Brachystegia spiciformis*, *B. Boehmii*, and *Julbernardia globiflora*. The three show a decrease of the concentration of nitrogen, potassium, and phosphorus and an increase in calcium, from the time of leaf initiation to the middle of the dry season. Magnesium, on the other hand, first increases and then decreases in the same period. Ernst sees in this diminution through the life cycle two processes in action. The first, most important during the rainy season, is the loss of elements by washing out; the second, operative from the end of the rainy season, is the translocation of nutrients from the leaves to perennial organs.

The first researchers to analyze the nutrient content of the epigeous herbaceous biomass of neotropical grasslands and savannas, or at least of some of its most characteristic species, were French and Chaparro (1960). They investigated several bromatological variables and mineral nutrients of four grasses introduced for forage in different places in Venezuela (*Panicum maximum*, *P. purpurascens*, *Digitaria decumbens*, and *Hyparrehenia rufa*), as well as the aerial biomass of 12 species of native savanna grasses (species of *Trachypogon*, *Axonopus*, *Aristida*, *Paspalum*, *Leptocoryphium*, *Andropogon*, and *Reimarochloa*). The samples were taken during the dry season, and the mean values obtained in both groups of species are depicted in table 20. Although the content of the four elements studied, nitrogen, potassium, calcium, and magnesium, was very low in both groups of grasses, it was twice as high in the introduced species than in the native ones.

In an analysis of the native grasses of Surinam, Dirven (1963) concluded that with the exception of the aquatic species, all the other grasses were deficient in nitrogen, phosphorus, and calcium, as well as zinc and copper. (His point of reference was the minimum requirements of dairy cows.)

Some herbaceous and woody species of the hyperseasonal savanna of Aripo, Trinidad, were analyzed by Ahmad and Jones (1969), who found values similar to the ones obtained by French and Chaparro for the Venezuelan species native to the seasonal savannas (see fig. 19). The Cyperaceae (*Lagenocarpus*, *Rhynchospora*, *Scleria*) had a nutrient content even lower than that of the grasses (*Panicum*, *Paspalum*, *Thrasya*), while various broad-leaved and woody species (*Byrsonima*, *Miconia*, *Sauvagesia*, *Chrysobalanus*, and so on) in general have higher concentrations.

Gomide and his fellow researchers (1969) analyzed a number of forage grasses of African origin that were introduced into central Brazil, where they were grown with and without nitrogen fertilizer. The data for the analysis of the plants grown without fertilizer are shown in table 20 and agree with the data for the same type of grasses from Venezuela.

The forage value and the nutrient content of 13 species of grasses of the hyperseasonal savannas and marshes of Apure (species of *Panicum, Luziola, Leersia, Hymenachne, Axonopus, Sporobolus,* and *Imperata*) were studied during the growing season by González Jiménez and Escobar (1975) and González Jiménez (1979). The average of the values for each nutrient are shown in table 20. Although the nutrient content is still very low, it is appreciably higher than that determined by French and Chaparro for the grasses of the savanna during the dry season.

Medina et al. (1977) analyzed the leaves and the aerial tissue of the two dominant species of the savanna of Calabozo, *Trachypogon vestitus* and *Axonopus canescens,* for six nutrients. These species had been subjected to treatments of fertilization, burning, and cutting. Data on the aerial portion subjected to fire, and on the 4 and 6-month harvest (average of both values) toward the end of the growing season, are shown in table 20. The results, especially for phosphorus and potassium, are considerably lower than the mean of those obtained for the Apure grasslands. On the other hand, the amount of nitrogen, phosphorus, and potassium increased significantly in the fertilized plots.

All the preceding values of mineral composition in one or another herbaceous species were obtained at a specific point in the annual cycle, whether at the beginning, in the middle, or toward the end, and consequently only a static view can be obtained. It is easy to surmise that nutrient values vary considerably in the course of the yearly cycle according to species and ecosystem, and it is therefore absolutely necessary to follow the data throughout the year, whether for individual species or for the entire herbaceous stratum of a savanna. González (1977) was first to investigate the amount of eight elements (phosphorus, calcium, magnesium, potassium, sodium, iron, copper, and zinc) present in two tropical grasses and the various bromatological indices for the same, from May to January, or practically throughout the growing season. These grasses are dominant in the savannas of Apure and very common in all neotropical savannas: *Axonopus purpusii* and *Leer-*

**Table 20.** Content of five nutrients in savanna biomass, grasses, and trees, cultivated forage plants, and some tropical crops.

| Type of vegetation | Nutrients | | | | | Observations | Sources |
|---|---|---|---|---|---|---|---|
| | N (%) | P (%) | K (%) | Ca (%) | Mg (%) | | |
| *Neotropical Savannas* | | | | | | | |
| 1 Savanna grasses | — | 0.06 | 0.52 | 0.11 | 0.27 | Mean of 16 samples of aerial biomass of 12 species. Dry season | French and Chaparro (1960) |
| 2 Dominant grasses and sedges in savannas | 0.71 | 0.02 | 0.32 | 0.13 | 0.10 | Aerial parts; mean of 5 grasses and 5 sedges. | Ahmad and Jones (1969) |
| 3 Savanna grasses | — | 0.16 | 1.39 | 0.20 | 0.24 | Aerial parts; mean of 13 species. Rainy season. | González Jiménez (1978) |
| 4 Dominant grasses in seasonal savannas | 0.81 | 0.07 | 0.51 | 0.18 | 0.16 | Aerial parts; mean of 2 species. Wet season. | Medina et al. (1977) |
| 5 Native grasses of forage value in the savanna | — | 0.15 | 0.40 | 0.16 | 0.17 | Aerial parts; mean of 2 species; minimum values. Dry season. | González (1977) |
| 6 Native grasses of forage value in the savanna | — | 0.50 | 2.12 | 0.59 | 0.47 | Aerial parts; mean of 2 species; maximum values. Wet season. | González (1977) |
| 7 Entire herbaceous stratum of the savanna | 0.34 | 0.06 | 0.35 | 0.19 | 0.17 | Aerial parts; mean of 3 communities; minimum values. Dry season. | Sarmiento (unpubl.) |
| 8 Entire herbaceous stratum of the savanna | 0.84 | 0.12 | 1.06 | 0.59 | 0.72 | Aerial parts; mean of 3 communities; maximum values. Wet season. | Sarmiento (unpubl.) |

| | | | | | | | |
|---|---|---|---|---|---|---|---|
| 9 | Entire herbaceous stratum of the savanna | 0.62 | 0.08 | 0.20 | 0.46 | 0.10 | Underground parts; mean of 3 communities. Wet season. | Sarmiento (unpubl.) |
| | Mean of 1 to 8 | 0.67 | 0.14 | 0.83 | 0.27 | 0.29 | | |

*Introduced Forage Grasses*

| | | | | | | | |
|---|---|---|---|---|---|---|---|
| 10 | African forage grasses in Venezuela | — | 0.12 | 1.43 | 0.22 | 0.37 | Aerial parts; mean of 4 species; 31 determinations. Dry season. | French and Chaparro (1960) |
| 11 | African forage grasses in Brazil | — | 0.26 | 1.42 | — | — | Aerial parts; mean of 6 species. Wet season. | Gomide et al. (1969) |

*Cultivated Species*

| | | | | | | | |
|---|---|---|---|---|---|---|---|
| 12 | *Medicago sativa* (alfalfa), Venezuela | — | 0.19 | 1.78 | 0.98 | 0.22 | Aerial parts. | French and Chaparro (1960) |
| 13 | *Oryza sativa* (rice), Trinidad | 2.69 | 0.30 | 2.20 | 0.37 | 0.19 | Aerial parts | Ahmad and Jones (1969) |
| 14 | *Ipomea batatas* (sweet potato), Trinidad | 3.20 | 0.25 | 1.88 | 1.34 | 0.54 | Leaves | Ahmad and Jones (1969) |
| | Mean of 12, 13, and 14 | 2.94 | 0.25 | 1.95 | 0.90 | 0.32 | | |

*Savanna Trees*

| | | | | | | | |
|---|---|---|---|---|---|---|---|
| 15 | Evergreen trees, Venezuela | 0.86 | 0.05 | 0.68 | 0.64 | 0.28 | Leaves, mean of 2 species | Montes and Medina (1975) |
| 16 | Deciduous trees, Venezuela | 1.74 | 0.11 | 2.16 | 0.89 | 0.49 | Leaves, mean of 2 species. | Montes and Medina (1975) |

*sia hexandra.* Table 20 shows the extreme values for both. The four elements included in table 20 (phosphorus, potassium, calcium, and magnesium), as well as the other four, exhibited the lowest values in January, well into the dry season, while the highest correspond in almost all cases to June-July, which is the time of most active growth. Although the minimum values were quite low, in many cases well below the minimum requirements for dairy cattle, the values reached during the growing season compared favorably with those of any well-managed tropical grazing land with introduced species.

Finally we will consider our own data on the nutrient content of the total aerial biomass of three savannas in the llanos of Barinas (see table 20). In addition to the studies of structure, production, and water balance described earlier, we determined the percentage of nitrogen, phosphorus, potassium, calcium, and magnesium in the biomass of the three. The first was a savanna grassland of *Axonopus*

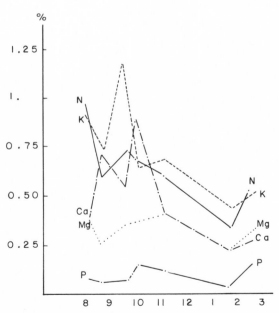

**Figure 34** Variation in the content of five nutrients per dry weight unit in the aerial biomass of a seasonal savanna of *Axonopus purpusii – Leptocoryphium lanatum* (Barinas), from the middle of the rainy season to the end of the dry season.

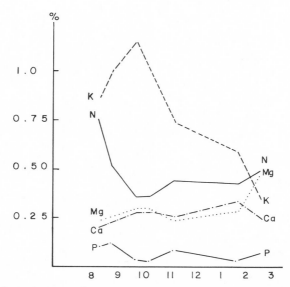

**Figure 35** Variation in the content of five nutrients per dry weight unit in the aerial biomass of a hyperseasonal savanna of *Sorghastrum parviflorum* (Jaboncillo), from the middle of the rainy season to the end of the dry one.

*purpusii, Leptocoryphium lanatum,* and *Trachypogon vestitus,* over well-drained alfisols (oxic paleoustalf, Barinas series); the second, a herbaceous community of *Sorghastrum parviflorum* over hydromorphic soils (typic tropaqualf, Jaboncillo); the third, an open savanna with a herbaceous stratum dominated by *Leptocoryphium lanatum* and *Elyonurus adustus* over even poorer soils (ultic haplustalf, Boconoito series). Figures 34, 35, and 36 represent the variations in those nutrients in each of the communities, from their peak vegetative phase in the middle of the humid season to the phase of maximum decline of the green biomass just before the new vegetative cycle starts with fire and rains at the end of the dry season. Table 20 only has the maximum and minimum values for the three communities in a block.

The results indicate that nitrogen, phosphorus, potassium, and calcium diminished in the aerial biomass as it dried out during the dry season, from the end of the reproductive cycle until the end of the dry period. The concentration of magnesium oscillated without any definite pattern. In the savanna of Jaboncillo, the reduction in concentration was more abrupt and took place earlier than in the

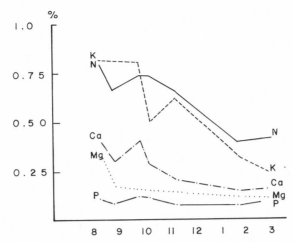

**Figure 36**  Variation in the content of five nutrients per dry weight unit in the aerial biomass of a seasonal savanna of *Leptocoryphium lanatum – Elyonurus adustus* (Boconoito), from the middle of the rainy season to the end of the dry one.

other two, reaching the minimum value already before the end of the rainy period but stabilizing from there on. This could be related to the more clear-cut seasonal contrasts of this hyperseasonal savanna, while the other two are seasonal savannas.

The extreme nutrient concentration values described in table 20 are very low in the dry season (except for magnesium). At their highest they approach the mean concentration of nitrogen, phosphorus, and calcium in xerophytic grasses, while potassium and magnesium reach even higher values. It should be pointed out that our measurements do not include the principal growing season, when, according to the pattern uncovered in other studies, the nutrients reach their highest values.

If instead of the percentages of nutrients in the biomass we consider their absolute quantities, at the same time that we take into account the continual increases of the total aerial biomass of the herbaceous stratum during the annual cycle (see chapter 4), we find that the nutrients keep increasing until the dry season begins, when they stabilize or decrease slightly. That is, during some time between the end of the reproductive stage and the beginning of the dry season, the increase in aerial biomass counterbalances the decrease in the concentration of mineral elements. Later, during the dry season, the reduction in concentration per unit of biomass is greater than the total increase in that biomass, resulting in an absolute reduction in mineral content.

We also performed two analyses of mineral concentration in underground biomass in the wet season (see table 20). The results indicate that at that time (September and October) potassium and magnesium had lower values in the underground tissues than in the above-ground portion, in each of the three savannas, while the concentrations of nitrogen, phosphorus and calcium were practically equivalent in both parts. However, at that time, as well as during most of the annual cycle, the underground biomass is larger than the aerial one, so the values of nutrients expressed in terms of kg/ha are greater for the hypogeous portion.

The findings on the nutrient content of the savanna vegetation may be summarized as follows:

1. The concentration of nitrogen and calcium in the aerial portion of grasses and other herbaceous plants is notably lower than the concentration of these elements in the leaves of tropical rain forests, whereas the two ecosystems are equivalent in their content of potassium, phosphorus, and magnesium. In general, the concentration of nitrogen in savanna species is only one-half that in forest species, and that of calcium, a third. This comparison may be thought of as slightly biased since it compares tree leaves with the total aerial portion of savanna species, but it holds even if only the leaves of grasses are analyzed.

2. The nutrient content in the aerial portion of savanna species is only half to a third that of some cultivated plants such as alfalfa (Leguminosae), rice (Gramineae), and sweet potato (Convolvulaceae). This applies to nitrogen, phosphorus, potassium, and calcium. There are no significant differences in the concentration of magnesium.

3. The trees of the savannas exhibit a very variable pattern. Some, especially the evergreen sclerophyllous species like *Curatella americana* and *Byrsonima crassifolia*, are poorer in nitrogen, phosphorus, potassium, and calcium than the average forest tree, while the deciduous species and those that are phylogenetically related to forest species (vicariants or in the same genus) are equivalent to

them with regard to nutrient accumulation, excepting perhaps for more potassium and less sodium. The concentration of nitrogen, phosphorus, and potassium diminishes during the wet season due to washing by rains, and during the dry season as a result of translocation.

4. The sclerophyllous trees of the savanna have more calcium and less phosphorus than the grasses; the latter in turn are richer in nitrogen, phosphorus, calcium, and potassium than the sedges.

5. During the period of most active growth, at the beginning of the rainy season, and in the final phase of decline during the dry season, the percentage of the five nutrients in the leaves is reduced to a half, to a third, or even less.

6. There may be great differences between grass species of the same or a different savanna community. The best native forage grasses have higher nutrient contents during the wet season than do the introduced species, while the introduced forage species have a higher nutrient content than the natives during the dry season, especially for potassium and calcium.

7. According to the scant available data, the hypogeous biomass of savanna plants contains a lower percentage of potassium and magnesium than does the aerial biomass, while the proportions of nitrogen, phosphorus, and calcium are similar.

8. On the average, the values of phosphorus and potassium (but not of calcium and magnesium) found in savanna grasses during the dry season are lower than the minimum required by cattle. During the wet season these species fulfill the nutrient requirements of mature beef cattle.

# The problem of aluminum in the savannas

Goodland (1971) described aluminum as an ecological factor with a strong negative influence over the cerrado vegetation, because its presence makes phosphorus and calcium insoluble. Consequently, not only would the cerrado species be oligotrophic-sclerophyllous because of the oligotrophic nature of soils, but also as a result of aluminum toxicity. That is, the sclerophyllous characteristics would indirectly be the result of the high concentration of aluminum in the soil.

The toxicity of aluminum is caused by two mechanisms. On the one hand there is a direct toxic effect, since this element thwarts the growth of roots by inhibiting the mechanisms of phosphorylation in the cells and thereby interfering with the mitotic cycle (Clarkson, 1969). Aluminum also has an indirect effect: in a very acid medium, in the presence of aluminum, phosphate ions precipitate as insoluble aluminum phosphate (Coleman et al., 1960) that cannot be utilized by the plants. In this respect aluminum is analogous in its action to iron. The toxicity of aluminum is blocked in certain acid-tolerant species that immobilize it through a nonspecific chelation mechanism, as iron may also be immobilized (Grime and Hodgson, 1969).

The ferralitic soils of the savannas can have important quantities of aluminum. Goodland (1971) found that in 110 superficial samples of cerrado soils from central Brazil, the mean aluminum quantity was 75 ppm, and values of 100 ppm were frequent. Eiten (1972) counted aluminum values of up to 3.5 and 4 meq/100 g (that is up to 360 ppm) on the basis of analyses of representative soils of the whole cerrado area, noting that forest soils, even the latosols, normally have less exchangeable aluminum than cerrado soils. Soares et al. (1974) examined three oxisols under cerrado in the area of Brasilia and found aluminum saturation values of 70 to 86%, while the norm of aluminum saturation for all cerrado soils is greater than 50%. Values of the same order are also frequent in the soils of the Colombian llanos under savanna vegetation (FAO, 1965). Kamprath (1972) studied 26 soils of the llanos of Colombia and measured a mean of exchangeable aluminum of 3.05 meq/100 g, with maximum values of 5.8 meq/100 g, the average aluminum saturation being 68%, with a maximum value of 86%.

In plants, aluminum is found in the proportion of 20 p.p.m./dry weight. However, there are many species that not only tolerate this element but accumulate it in large quantities, with values up to 1000 p.p.m./dry weight. These species are not necessarily phylogenetically related, although there are families such as Vochysiaceae, Rubiaceae, Symplocaceae, and Melastomataceae with many accumulating species, while other families such as Leguminosae have none. Apparently in the tolerant species there is a mechanism as yet not well understood that impedes the entrance of aluminum into the cells, the element being precipitated in the cell walls (Clarkson, 1969).

The analysis conducted by Ahmad and Jones (1969) shows that the concentration of aluminum in the aerial parts of grasses and sedges in the savanna of Aripo (Trinidad) reach very high values. On the basis of their data the mean for 10 species is 912 p.p.m., with maximum values of 4040 p.p.m., in *Panicum stenoides*. However, the single maximum in this savanna is found in a herbaceous Melastomataceae, *Acisanthera uniflora*, with a concentration of 20,200 p.p.m.! This species is also common in the Venezuelan savannas. In comparison with these values, Ahmad and Jones point out that the concentration of aluminum in rice is 40 p.p.m. and in sweet potatoes it cannot be detected. The exchangeable aluminum in the soil of the savanna of Aripo is also very high, varying between 0.22 meq/100 g in the 0–18 cm horizon to 9.9 meq /100 g in the 60–250 cm horizon. These authors note that the high concentration of aluminum in *Panicum stenoides* is associated with very low values of phosphorus, potassium, sulfur, and iron, lower even than in those of other species of the same community, which are very low already.

Medina (1978) considers that the toxicity of aluminum in savannas has not been demonstrated and maintains on the basis of the data of Gottlieb et al. (1966) that there are no significant differences in the aluminum concentration of savanna and forest species, and that the differences are related more to taxonomic groups than to ecosystems. To be sure, taxonomic relationships must be taken into account, as for example the fact that Leguminosae do not accumulate aluminum while the Vochysaceae have many species that do. However, an analysis of Gottlieb's data discloses that the mean aluminum concentration in cerrado species is 53 p.p.m. and in forest trees, 33 p.p.m. Both of these are relatively low values owing to the large number of Leguminosae in the samples. Moreover, three species, all from the cerrado, reach values above 100 p.p.m. of aluminum.

The question of aluminum in relation to scleromorphism and distribution in the savannas has not yet been solved, but some generalizations may be made. In the first place we must recognize that ferralitic soils, by the very nature of the process of ferralization, show high concentration of $Al^{+++}$ in the exchange complex. This cation can become the principal exchangeable cation, more abundant than $H^+$ and of course much more so than any other metal ion, and even more than the sum of all metallic ions. This happens in very evolved ferralitic soils, in both forests and savannas.

On the other hand, the direct toxicity of aluminum has been well established and the cellular mechanism outlined. It is also clear that in a very acid medium aluminum has an indirect toxic effect in precipitating soluble phosphorus into insoluble aluminum phosphates. In this case the deficiency of phosphorus could be the cause of the foliar scleromorphism of the woody species, in the same way as Loveless (1961, 1962) described phosphorus deficiencies in other tropical ecosystems. It is also possible that calcium may have an effect on the toxicity of aluminum; one study showed that an aluminum-sensitive soybean variety was adversely affected by decreasing calcium levels considerably more than a resistant variety (Lunt 1972).

In tropical forests, whether on podzols or on latosols, the principal nutrient cycle is presumed to go directly from the litter to the vegetation without passing through the mineral soil (Went and Stark, 1968; Herrera et al., 1978). In the savannas it is most probably an open cycle, in the sense that absorption takes place directly from the mineral soil. For this reason, even if aluminum were present in the same concentrations in forest and savanna soils, it should be less toxic to trees that shortcircuit the mineral cycle and absorb very little from the mineral content of the soil, than to savanna species, which obtain their nutrients from the soil solution in equilibrium with an exchange complex dominated by the aluminum ion.

In the savannas, both the herbaceous and the woody species show a clear tendency to accumulate aluminum above the mean of all species, and many of them, in both life forms, are notable accumulators. It is not yet clear what this capacity of savanna species means, either in terms of the possibility of ecological success in a medium rich in aluminum, or in their ecomorphological characteristics, caused indirectly by this element. But the aluminum factor must be taken into consideration whenever the adaptive mechanisms of the species to the conditions of tropical savannas are discussed.

# Other nutrients and mineral elements in the vegetation

There is very little information regarding the presence of sulphur in the savanna. From the data of Ahmad and Jones (1969) for the savanna of Aripo (Trinidad) we calculated the average sulphur content in 10 species (5 grasses, 5 sedges) to be 0.06%, with a range of variation among the different species of 0.01 to 0.11%. On the other hand, the tree *Byrsonima crassifolia* had a concentration of 0.04% in its leaves.

Medina et al. (1977) calculated the content of sulphur in the aerial biomass of two grass species in the savannas of Calabozo: *Trachypogon plumosus* and *Axonopus canescens*. They found that it varies throughout the year in parallel form in both species. The maximum value found at the beginning of the growing season, in May, reached values of 0.6 to 0.7% of the aerial dry weight, while in November, at the beginning of the dry season, it had diminished to 0.1 – 0.2%, and in January it only reached 0.08% in the two species. The fall in the value of sulphur in the aerial biomass of the savannas during the dry season is very steep. At the time of burning the amount of sulphur in the aerial portion is in the order of 5 kg/ha (assuming a content of 0.1% and a biomass of 500 g/m²). Although all of this is probably lost due to volatilization during burning, it only represents a value of the same order of magnitude as the normal presence of this element in rainwater, and much lower than what rains carry in areas with urban or industrial influence.

The existing data give us no reason to suspect any deficiency or toxicity of iron, copper, zinc, boron, and chlorine in the savannas. It is difficult to imagine deficiencies of iron or manganese in acid soils like these; the content determined in different herbaceous species (Ahmad and Jones, 1969; González, 1977) oscillates between 30 and 300 p.p.m. for iron (with a maximum of almost 1000 p.p.m. for a sedge from Trinidad) and between 5 and 25 p.p.m. for manganese. This last element can, however, become quite toxic in a very acid and anaerobic medium, and several crops in the cerrado are sensitive to its action (Malavolta et al., 1977).

In the leaves of forest trees Stark (1970) measured 50 p.p.m. of iron. The same scientist obtained copper values of 7 to 12 p.p.m. in forest trees, while González Jiménez and Escobar (1975) found for this same element a range of variation in the grasses of the savannas of Apure of 8 to 48 p.p.m. On the other hand, in two species of grasses of the same savanna the annual variation oscillated between 10 and 100 p.p.m. (González, 1977). For zinc the same study indicated an annual variation between 20 and 220 p.p.m. and an interspecific range of 65 to 225 p.p.m.

Although we have no information regarding chlorine, this element is widely distributed in the soils and no deficiency or excess in this element is to be expected; the same applies to sodium. As far as boron, Gottlieb et al. (1966) find up to 3 ppm in leaves of forest and savanna trees, with no apparent differences between these two groups and no apparent toxicity or deficiency.

Finally we must consider the case of molybdenum, since deficiences in this micronutrient can occur under certain edaphic situations. We possess no information about the content of molybdenum in savanna plants, but deficiencies can be expected because molybdenum behaves in a manner similar to phosphorus: in an acid environment insoluble molybdenum salts of iron and aluminum are formed. Williams et al. (1972) were unable to detect the presence of molybdenum in the exchange complex in three Amazonian soils, one an upland forest, another a flooded forest, and the last an agricultural plot within the first forest.

Last to be mentioned is silica ($SiO_2$), the most abundant mineral in the ash of savanna species, where it can reach values of 5% (Egunjobi, 1972) and even 10% in savannas of *Trachypogon* at the end of the dry season (Espinoza, 1969). It is important to remember these figures when precise calculations of production are made, in order to subtract them from the mean biomass in the calculation of the real amount of organic material produced by the vegetation.

# The nutrient cycles

## Comparison of the pool of soil and vegetation nutrients

A quantitative picture of the nutritional factor of neotropical savannas should relate the nutrient content in the soil to that in the vegetation, as well as to the quantities accumulated during an annual production cycle. On the basis of the available data, as a first approximation, we will quantify the principal transfer functions between compartments in a simple model of the nutrient economy. This will make it possible to single out the critical elements that condition the success of a given plant formation in a specific habitat within the wet tropics, and to determine whether nutrients really play an important role in favoring a given ecosystem over other alternatives.

The data for the savannas will be the same that we have already analyzed previously in relation to structure, production, and water balance. These are the savannas of Barinas, Boconoito, and Jaboncillo, the first two seasonal, the last hyperseasonal. We then will compare these results with some partial results about rain forest ecosystems in the western llanos and with the more complete data of other researchers in different tropical rain forests. The mineral cycles in tropical rain forests are actually better known than those in herbaceous communities, and it is frequently necessary to extrapolate their results to the case of the savannas.

Table 21 shows the content of the five nutrients in the soils of the three savannas. In the case of nitrogen we are dealing with total content; later on we will see what information can be obtained regarding the nitrogen available to the plant. The value for phosphorus corresponds to soluble phosphorus; calcium, magnesium, and potassium correspond to quantities in the exchange complex. To carry these data to g/m² we have taken into account all the soil horizons up to 100 cm in depth, since this level contains 70% of the hypogeous biomass in the savanna of Barinas, 97% in Boconoito, and 99.4% in Jaboncillo. In table 22 we present similar calculations using the soil depth where 95% of the hypogeous biomass of each community is found, which in Barinas resulted in a depth of 160 cm, in Boconoito 80 cm, and in Jaboncillo only 40 cm. The first set of data allow a more uniform comparison from an edaphic point of view, the second one a more ecological analysis on the assumption that a high proportion of the nutrients come from that part of the soil where most of the roots are found. Thus, for example, in soils with a hydromorphic, impermeable, and extremely compact horizon at 40 cm (the savanna of Jaboncillo), the nutrients below this horizon are practically beyond the reach of the herbaceous species.

Table 23 provides the data about nutrients in the vegetation. The quantities of nutrients are measured by unit of surface in both the hypogeous and epigeous biomass, using in each case the highest values seen during several annual cycles, since they are the ones that offer information about maximum accumulation.

**Table 21.** Nutrients by surface unit (g/m²) in the first 100 cm of soil, in three savanna ecosystems in the western llanos of Venezuela.

| Location | Total N | Soluble P | Interchangeable Ca | Interchangeable Mg | Interchangeable K |
|----------|---------|-----------|--------------------|--------------------|--------------------|
| Barinas | 586 | 1.12 | 101 | 58 | 71 |
| Jaboncillo | 597 | 1.39 | 1101 | 641 | 71 |
| Boconoito | 531 | 2.12 | 23 | 13 | 31 |

**Table 22.** Nutrients per surface unit (g/m²) in the soil profile containing 95% of the underground biomass, in three savanna ecosystems in the western llanos of Venezuela.

| Location and depth | Total N | Soluble P | Exchangeable Ca | Exchangeable Mg | Exchangeable K |
|---|---|---|---|---|---|
| Barinas 0–160 cm | 790 | 1.83 | 254 | 188 | 135 |
| Jaboncillo 0–40 cm | 293 | 0.72 | 32 | 15 | 16 |
| Boconoíto 0–80 cm | 499 | 1.87 | 20 | 11 | 26 |

Unfortunately we do not have precise biomass data or information about nutrient contents in the two types of rain forests in the western llanos that we have studied from the perspectives of composition of soils and water balance: the forest gallery of the large rivers and the semideciduous forests of the Andean foothills. Both communities have regional extensions interdigitating with the savannas to form various vegetational mosaics.

To supplement the data for the western llanos, we have used two of the most exhaustive studies done to date in low altitude neotropical forests: one by Golley et al. (1975) on the semideciduous forest of Darien (Panama), and the other by Hase and Folster (1982) on the Caparo forest in Venezuela. The moist tropical forest in Panamá, recognized by Golley and his colleagues as a transitional type between a typical moist tropical forest and a dry tropical forest, has many similarities with the forests of the more humid areas of the llanos of Colombia and Venezuela. The climatic conditions in which both kinds of ecosystems prosper are very similar, since the forests in Panamá receive 1900 to 2000 mm of yearly rainfall, but with three dry months of rainfall below 50 mm (January to March) and April having barely above 50 mm — a climate quite comparable to that of the forests of the llanos. The floristic similarities are also high, since with one or two exceptions, the dominant species are the same. The Panamanian forest is found in soil that is relatively rich in nutrients for the wet tropics, owing to the high content of montmorillonite that leads to cation exchange capacities of up to 50 meq/100 g and a base saturation of 86%. However, the young soils of the Venezuelan and Colombian llanos

**Table 23.** Nutrients per surface unit (g/m²) in the epigeous (E) and hypogeous (H) biomass of three savanna ecosystems of the llanos of Venezuela.

| Location | Nitrogen (E) | Nitrogen (H) | Phosphorus (E) | Phosphorus (H) | Calcium (E) | Calcium (H) | Magnesium (E) | Magnesium (H) | Potassium (E) | Potassium (H) |
|---|---|---|---|---|---|---|---|---|---|---|
| Barinas | 6 | 4 | 1.16 | 0.7 | 7 | 4 | 3 | 1 | 6 | 2 |
| Jaboncillo | 6 | 6 | 1.42 | 0.7 | 6 | 3 | 2 | 1 | 12 | 2 |
| Boconoito | 3 | 9 | 0.48 | 1.68 | 2 | 3 | 3 | 1 | 3 | 3 |

(inceptisols) are also relatively rich in nutrients, as can be appreciated in the values of the soil profile of the Q1 terrace at the Santo Domingo River (table 24). Consequently, a comparison of the nutrient economy of the Panamanian forest and the Venezuelan savannas evaluates two ecosystems that grow under relatively similar environmental conditions.

The basic results of the analysis of the mineral elements in the vegetation and soil of the semideciduous forest of Darien are summarized in table 25. This study did not include nitrogen, only phosphorus, potassium, calcium, and magnesium.

The other forest ecosystem is the semi-evergreen seasonal forest of the Caparo Forest Reserve, in the western llanos of Venezuela (Hase and Folster, 1982). The above-ground biomass, which grows on eutrophic, well-drained alluvial soils, attains 402 tons $\cdot$ ha$^{-1}$. The distribution of five nutrients in the biomass and the soil (0 – 50 cm) is shown in table 26. After discussing the nutrient economy of savanna ecosystems, we will return to these data to compare the nutrient balance of tropical forest and savannas.

Lamotte and Bourlière (1983), in the latest survey of nutrient cycling in tropical savannas, point out the scarcity of available information. These studies are still in their infancy, and their results must be treated with caution. Nitrogen and phosphorus have received some attention but next to nothing is known about the other elements. Keeping these remarks in mind, we present our own data with the limited objective of disclosing some major trends distinguishing savannas from rain forests.

The quantities of each nutrient in the soil and in the total biomass are compared in the three savannas (table 27). We obtained the relation of soil to vegetation by taking into account the total biomass as well as only the epigeous part, which is almost completely consumed by fire. Several interesting conclusions regarding the nutrient economy in savanna ecosystems and their difference with the rain forest ecosystems can be drawn. In our discussion the two seasonal systems (Barinas and Boconoito) will be separated from the hyperseasonal savanna of Jaboncillo.

In the seasonal savannas phosphorus can be divided into two equal portions, one in the soil and the other in the vegetation, both systems following the same pattern. But there are differences with regard to calcium, magnesium, and potassium in the poorest soils (Boconoito) as compared to the less poor ones (Barinas). The respective values for the available nutrients in the soil and the nutrients

**Table 24.** Nutrients per unit surface (kg/ha) in the soil of two forests in the western llanos of Venezuela at two depths (in cm).

| Type of forest | Total N | | Soluble P | | Exchangeable Ca | | Exchangeable Mg | | Exchangeable K | |
|---|---|---|---|---|---|---|---|---|---|---|
| | 0 – 30 | 0 – 100 | 0 – 30 | 0 – 100 | 0 – 30 | 0 – 100 | 0 – 30 | 0 – 100 | 0 – 30 | 0 – 100 |
| Gallery (on entisol) | 8310 | 15,210 | 26.2 | 48.6 | 12,020 | 27,370 | 3910 | 12,970 | 820 | 1570 |
| Low, semideciduous (on alfisol) | 1710 | 4520 | 15.5 | 36.8 | 50 | 150 | 30 | 70 | 210 | 600 |

**Table 25.** Nutrients per unit surface (kg/ha) in two semideciduous parcels in Darien (Panama), and ratio of nutrients in vegetation ($V$) to vegetation plus soil ($S$). Vegetation includes the total of all plant organs; soil depth is the first 30 cm (data from Golley et al., 1975).

| | Rio Lara | | | Rio Sabana | | |
|---|---|---|---|---|---|---|
| | Vegetation | Soil | % $V/(V + S)$ | Vegetation | Soil | % $V/(V + S)$ |
| Phosphorus | 241 | 33 | 88 | 85 | 11 | 89 |
| Potassium | 4598 | 508 | 90 | 1606 | 197 | 89 |
| Calcium | 4702 | 18,582 | 20 | 3502 | 25,749 | 12 |
| Magnesium | 437 | 2830 | 13 | 423 | 2281 | 16 |

**Table 26.** Nutrients per unit surface (kg/ha) and ratio of nutrients in vegetation (V) to vegetation plus soil (S) in the semi-evergreen seasonal forest in the Caparo Forest Reserve, Venezuela (data from Hase and Fölster, 1982).

| | Aboveground biomass | Soil (0 – 50 cm) | $V/(V+S)$ (%) |
|---|---|---|---|
| Nitrogen | 1980 | 4071 | 32.8 |
| Phosphorus | 290 | 2484 | 10.5 |
| Potassium | 1820 | 419 | 81.3 |
| Calcium | 3380 | 1858 | 64.5 |
| Magnesium | 310 | 423 | 42.5 |

accumulated in the total biomass are 23 and 4 for calcium; 47 and 3 for magnesium; and 17 and 4 for potassium. That is, the soils have quantities of these elements several times higher than the amount accumulated in the biomass.

It is therefore evident that phosphorus is the critical element in the development of the vegetation and in the primary production of these seasonal savannas. Although calcium, magnesium, and potassium occur in the respective soils in amounts several times greater than their concentration in the vegetation, they can go down to critical levels in poor soils such as those of Boconoito, or even poorer. In the hyperseasonal savanna, phosphorus is two to three times more abundant in the vegetation than in the soil; potassium is equally divided between the biomass and the soil exchange complex; calcium and magnesium are four to five times more abundant in the soil than in the vegetation. The nutrient balance is even more critical in this type of ecosystem than in the seasonal savanna, with phosphorus and potassium very limiting while calcium and magnesium have the potential of becoming limiting in poorer soils or in the case of massive export of nutrients from the system through harvest or consumption.

# The nitrogen economy

Nitrogen is a special case. It is found in soils at a concentration that is 24 to 80 times greater than that in the biomass of the three savannas. However, only a small fraction of the soil nitrogen is in a form that is

**Table 27.** Percentage of each nutrient accumulated in the total biomass and the aerial biomass at the time of greatest development in three savannas of the western llanos of Venezuela. The first depth is 100 cm; the second is where 95% of the hypogeous vegetation was found.

| Savanna type and location | Soil depths (cm) | N | | P | | Ca | | Mg | | K | |
|---|---|---|---|---|---|---|---|---|---|---|---|
| | | Total | Epigeous | Total | Epigeous | Total | Epigeous | Total | Epigeous | Total | Epigeous |
| Seasonal (Barinas) | 0–100 | 2.6 | 1.6 | 62.4 | 38.9 | 9.0 | 5.7 | 6.4 | 4.8 | 10.1 | 7.6 |
| | 0–160 | 1.9 | 1.2 | 50.4 | 31.4 | 3.7 | 2.6 | 2.1 | 1.6 | 5.6 | 4.2 |
| Hyperseasonal (Jaboncillo) | 0–100 | 1.9 | 1.0 | 60.4 | 40.4 | 0.8 | 0.5 | 0.5 | 0.3 | 16.4 | 14.1 |
| | 0–40 | 4.0 | 2.0 | 74.6 | 50.0 | 22.0 | 14.6 | 16.7 | 11.1 | 46.7 | 40.0 |
| Seasonal (Boconoito) | 0–100 | 2.2 | 1.7 | 50.4 | 11.2 | 17.9 | 7.1 | 23.5 | 17.6 | 16.2 | 8.1 |
| | 0–80 | 2.3 | 1.7 | 53.6 | 11.9 | 20.0 | 8.0 | 26.7 | 20.0 | 18.7 | 9.4 |

accessible to the vegetation. In these three savannas the nitrogen accumulated in the biomass is 10 to 12 g/m² (100 to 120 kg · ha⁻¹), of which almost half is renewed annually in aerial and underground production. The question then arises whether there is enough available nitrogen in the soil, or whether on the contrary it has all been used, and therefore has become an important critical element for the productivity of these ecosystems. To answer this question we will use the ecosystem model represented in figure 37, which has five fundamental compartments: live epigeous biomass, dry epigeous biomass, hypogeous biomass, organic soil nitrogen, and mineral soil nitrogen. The atmosphere is depicted as a reservoir of gaseous nitrogen, while the soil microorganisms act as the engines of a process than transfers nitrogen from one subsystem to another.

The concrete example will be the seasonal savanna of *Axonopus purpusii-Leptocoryphium lanatum* (Barinas), where the live epigeous biomass of the vegetation at its maximum development accumulated 60 kg · ha⁻¹ of nitrogen (see fig. 37). As the aerial organs dry out during the dry season, an important portion of this nitrogen is translocated toward the perennial underground organs, while another portion remains in the dry standing biomass. According to a study done on the recovery of various elements in the annual closed cycle of nutrients within the green biomass (Medina et al., 1977), about 66% of the nitrogen is saved by the plants through translocation. In the case of our savanna this means that 40 kg · ha⁻¹ would be translocated to the hypogeous organs while only 20 kg · ha⁻¹ would remain in the dry standing biomass. Most of the nitrogen that remains in the dry biomass will be lost by volatilization at the time of burning; however, a fraction is removed and returned by rainwater. As the dry biomass starts to be quantitatively important within the total epigeous biomass toward the middle of the wet season, it is subjected, as much as the green biomass, to a more or less intensive process of throughfall. Consequently our tentative assessment is that 25% of the nitrogen of the dry biomass is directly incorporated into the mineral nitrogen pool of the soil by throughfall. It is reasonable to expect that some 5 kg · ha⁻¹ are incorporated into the soil by this route before the end of the rainy season; the remaining 15 kg · ha⁻¹ are lost by volatilization.

On the other hand, the hypogeous biomass accumulates at its maximum development point some 40 kg · ha⁻¹ of nitrogen. Applying what is known about recycling time of the hypogeous

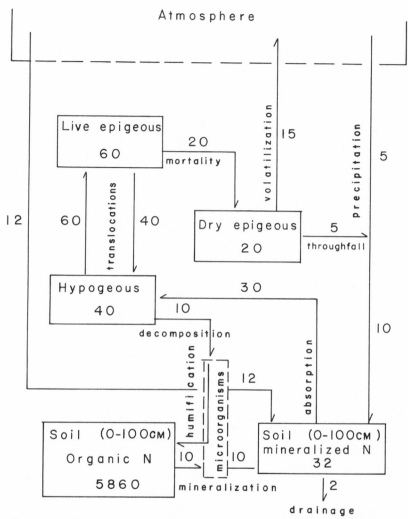

**Figure 37**  The annual nitrogen cycle in the seasonal savanna of *Axon-opus purpusii – Leptocoryphium lanatum* (Barinas). The values corresponding to the stocks accumulated in each compartment are given in tm.ha⁻¹. The values corresponding to the transfers between subsystems are given in tm.ha⁻¹. year⁻¹.

biomass in perennial grasses (see chapter 4), we have assumed that some 25% of the underground biomass is recycled every year through hypogeous decomposition. Consequently 10 kg · ha⁻¹ of nitrogen will pass through the soil microflora and microfauna and through the process of humification to the organic nitrogen pool of the soil. In addition, the 60 kg · ha$^{-1}$ that will form the epigeous biomass every year must pass through this soil compartment, which implies that the roots will absorb annually 30 kg · ha$^{-1}$ of mineral nitrogen from the soil, of which 20 kg · ha$^{-1}$, plus the 40 kg · ha$^{-1}$ translocated previously from the shoots, will be directed towards the aerial organs and the remaining 10 will remain in the underground organs, provided that the hypogeous biomass remains constant through time, as it should be in an ecosystem in equilibrium.

As the vegetation absorbs annually 30 kg · ha$^{-1}$ of mineral nitrogen from the soil, it is important to analyze the contributions that balance the inflow and outflow from this compartment. One is the rainfall contribution, which can be subdivided into two parts: one corresponds to the nitrogen incorporated directly by rainfall, the other is the amount of this mineral that is incorporated into the pool as a result of throughfall. We already estimated this quantity at 5 kg · ha$^{-1}$. With respect to the concentration of nitrogen in rainwater, Vera (1978) obtained for the region of the western llanos mean concentrations of nitrogen throughout the year of 0.48 p.p.m., which results in a contribution of 5 kg · ha$^{-1}$ · year$^{-1}$ for 1300 mm of annual rainfall. Consequently the total contribution of nitrogen through rainfall would be 10 kg · ha$^{-1}$ · year$^{-1}$.

The most important pool of nitrogen in this ecosystem is the organic nitrogen accumulated in the humus, which has a value of 5860 kg · ha$^{-1}$ in the horizon between 0 and 100 cm. If 10 kg · ha$^{-1}$ are incorporated from the decomposition of the root biomass that is recycled every year, in order to maintain equilibrium the same quantity must be mineralized each year; that is to say, it must pass from the more or less stable form of organic nitrogen in humus to the more reactive form of mineral nitrogen, through the processes of ammonia liberation and nitrification. This value is close to the figure given by Rham (1973) on nitrogen mineralization in a tropical savanna. In the seasonal savanna of Lamto (Ivory Coast) Rham found a yearly mineralization of 5 kg · ha$^{-1}$, an amount much lower than the mineralization occurring in a nearby tropical rain forest. The low mineralization appears to be due to the shortage of nitrifying bacteria in savanna soils, even during the rainy season (Meiklejohn, 1962).

Another interesting observation concerning mineralization of nitrogen in the soil is that there is a significant difference between the open grass savanna and the soil below trees. Rham noticed it in the small forest groves dispersed in the Lamto savanna (1973), and it was quantified by Bernhard-Reversat (1982) in the dry savannas of Senegal, where mineral nitrogen produced below isolated trees was twice the amount produced in the open.

Mineralization and rainfall alone contribute up to 20 kg $\cdot$ ha$^{-1}$ to the mineral system of the soils. But additional sources must be found in order to balance the 30 kg $\cdot$ ha$^{-1}$ that are absorbed every year by the vegetation. Moreover, there are other sources of loss from this labile mineral pool. A certain amount will be lost through drainage or deep infiltration, as well as through surface flow, but the loss of water in this manner is normally very small, amounting to no more than 150 to 200 mm a year. With nitrogen concentration in drainage and river water in tropical America being lower than 1 p.p.m. (Sioli, 1967, 1975; McColl, 1977; Klinge et al., 1977), we have calculated then that the losses through infiltration are about 2 kg $\cdot$ ha$^{-1}$ $\cdot$ year$^{-1}$.

Another process that may produce losses of nitrogen from the soil toward the atmosphere is bacterial denitrification, important in anaerobic microenvironments, but of negligible importance in the well-drained soils of these savannas. In summary, the losses of mineral nitrogen may be estimated at about 32 kg $\cdot$ ha$^{-1}$, as against 20 kg $\cdot$ ha$^{-1}$ of input, leaving 12 kg $\cdot$ ha$^{-1}$ still unaccounted for.

The only process that can be responsible for this net gain is the fixation of atmospheric nitrogen by means of free-living or symbiotic microorganisms. Dobereiner and Day (1974) maintain that it is the rhizosphere of the tropical grasses where the greatest nitrogen fixation by free-living bacteria of the genus *Beijerinckia* takes place. Under favorable conditions of humidity during the growing season, these bacteria can fix up to 1.5 kg $\cdot$ ha$^{-1}$ $\cdot$ day$^{-1}$ for a radical biomass of 5 tm $\cdot$ ha$^{-1}$. *Beijerinckia*, according to Moureaux and Bosquel (1973), is found in most ferralitic soils, even in the poorest, the rate of fixation being more active in the measure that the soils are poorer in nitrogen, so that the lower the nitrificating power of the soil the greater the bacterial fixation. It has already been pointed out that nitrification in tropical savanna soils appears to be insignificant, from which it can be inferred that the free fixation of nitrogen should be important. The same behavior is exhibited by the anaerobic fixers of the genus *Clostridium*, which, like *Beijerinckia*, become more abundant and active in the soils that are poorest in humus, carbon, nitrogen, and phosphorus.

Dobereiner (1979) maintains that the most important nitrogen-fixing association in tropical conditions is that between grasses and bacteria of the genus *Azospirillum*. These bacteria are found in most tropical soils with natural or cultivated Gramineae. They infect the roots, multiplying in the intercellular spaces. Under certain conditions of oxygen pressure, neither very high nor very low, these bacteria can utilize atmospheric nitrogen for protein synthesis.

There is no precise information on the total amount of nitrogen fixed in tropical soils by free soil microorganisms under herbaceous vegetation. Dobereiner and Day (1974) give a range of 50 to 1000 g $N \cdot ha^{-1} \cdot day^{-1}$ during the growing season for eight genera of tropical grasses common in African savannas, which have been cultivated or introduced in America. If we take the relatively low value of 100 g of $N \cdot ha^{-1} \cdot day^{-1}$ for a six-month growing season, we obtain a total amount fixed by these bacteria of $18 \text{ kg} \cdot ha^{-1} \cdot year^{-1}$. In latosols under grasses in Nigeria Moore (1963) indicates an annual fixation rate of $150 \text{ kg} \cdot ha^{-1}$, due primarily to *Clostridium* and other microorganisms that are not Azobacteriaceae. Moureaux and Bosquel (1973) give values of $50 \text{ kg} \cdot ha^{-1} \cdot year^{-1}$ as possible for nitrogen fixation by *Clostridium* in anaerobic conditions. In the seasonal savannas of Venezuela, especially in Barinas, the Cyanophyceae form an almost continuous crust on the soil, while the habitat of the hyperseasonal savannas, such as those of Jaboncillo, is particularly suited for the anaerobic fixation by *Clostridium* on account of its physical and chemical properties.

In Lamto (Ivory Coast) the soil of the savanna fixes appreciable quantities of atmospheric nitrogen through microorganisms associated with the roots of grasses (Balandreau, 1976). The hyperseasonal savanna dominated by *Loudetia* had an annual fixation of nitrogen of $13 \text{ kg} \cdot ha^{-1}$, and the seasonal savanna of *Hyparrhenia* had $9 \text{ kg} \cdot ha^{-1}$. On the other hand this author maintains that legumes do not play a major role in the few African savannas studied.

In contrast to the free fixation of nitrogen, the symbiotic fixation is quantitatively less important. The seasonal savannas of Venezuela grow a large number of Leguminosae that have well developed nodules (Barrios and González, 1971), but their quantitative contribution to the total biomass of the vegetation is rather modest. In the hyperseasonal savanna of the Venezuelan llanos the legumes may be completely absent (Sarmiento and Vera, 1978).

To balance our nitrogen account for the savanna under study, it is reasonable to adopt the working hypothesis that the 12 kg · ha$^{-1}$ of mineral nitrogen that were missing are contributed yearly by the free or symbiotic fixers. This completes the quantification of all the fundamental processes for the interchanges of nitrogen within the ecosystem. This balance does not account for the nitrogen accumulated in the primary or secondary consumers, especially because in the ecosystem of the tropical savanna the principal consumer is fire. Neither have we tried to calculate the maximum amount of nitrogen accumulated at the time of maximum development of the microflora. However, if in an ecosystem of temperate grasslands there are 2 tm · ha$^{-1}$ of dry weight of bacteria and Actinomycetes in the soil (Clark and Paul, 1970), the maximum quantity of nitrogen immobilized in this compartment of the savanna ecosystem must be of the same order, or some 20 kg · ha$^{-1}$, since the quantity of substrate available to heterotrophs in this tropical savanna is about the same as in the temperate grassland.

In summary, it is not reasonable to expect to find under savanna soils quantities of mineral nitrogen greater than what the vegetation incorporates during its annual growth, with the seasonal oscillations in the soils and vegetation derived from the annual rhythms of the productive processes in these ecosystems. Thus even the limited decomposition of the dead biomass that takes place during the dry period brings with it an increase in the amount of ammonium ions in the soil, since during this period there is no nitrification. On the other hand, at the beginning of the rainy season the ammonium disappears quickly and gives way to a liberation of nitrates that will soon be absorbed by the growing vegetation. However, independently of the vegetation cycles, we maintain that nitrogen is indeed one of the most important critical elements that control production in these savannas.

A confirmation of this hypothesis is obtained by analyzing the behavior of the ecosystem and its dominant species in response to the addition of different fertilizers. Medina and others (1977) fertilized two parcels of a savanna of *Trachypogon* in the llanos of Calabozo, one with nitrogen (67 kg · ha$^{-1}$) and the other with phosphorus (79 kg · ha$^{-1}$). After 256 days the epigeous biomass increased 30% in the parcel fertilized with nitrogen, and 18% in the one fertilized with phosphorus (in both cases potassium was also incorporated as a cation in quantities of 186 and 100 kg $\cdot$ ha$^{-1}$ respectively). The differences between the fertilized parcels and the control, burned but not fertilized, were significant at the 5% level. Likewise, the content of nitrogen and phosphorus in the aerial biomass of the two dominant grasses was significantly higher in comparison with the control in the respective treatments with fertilizers. The authors conclude that in these savannas nitrogen is the principal limiting element, although it is possible that in order to demonstrate more clearly its limiting effect it is necessary to apply a mix of nitrogen and phosphorus, since both are critical.

In the nitrogen economy of the seasonal savanna, the soil constitutes the principal reservoir of this element within the ecosystem. Even though the soils are relatively poor in organic matter, the pool of organic nitrogen represents a reservoir 40 or 50 times greater than the nitrogen accumulated in the rest of the system. Hence the deep and well-drained soils of the Barinas series (oxic paleoustalf), with a content of organic matter in the A1 horizon of 1 to 3%, have almost 6 tm $\cdot$ ha$^{-1}$ of nitrogen, compared to less than 200 kg $\cdot$ ha$^{-1}$ as a maximum for the rest of the vegetation.

In the second place, it is important to point out that the savanna recycles within its biomass a very high proportion (40% in this case) of the nitrogen that is utilized each year. This internal cycle, which bypasses the soil, had already been suggested by Ellemberg (1971). The remaining nitrogen comes in equal parts from three sources: throughfall, fixation of gaseous nitrogen, and mineralization of humus. As far as the losses of this element, the most important one takes place by volatilization to the atmosphere at the time of burning.

Finally, the rates of humification are rather modest and the time required to reconstitute the pool of organic nitrogen in the soil must be quite long, even if fires were less frequent and the nitrogen of the standing dry biomass and the litter could be incorporated into the soil through decomposition. Since the processes of humification and of mineralization of the humus take place simultaneously in the soil even before a stable equilibrium in the nitrogen cycle is reached, it can be easily seen that to form a nitrogen pool like the existing one can easily take hundreds of years.

It is interesting to compare, even if only qualitatively, the nitrogen cycle in the savanna subjected to fire to that in protected or only infrequently burned ones. In this last case the dead standing biomass accumulates until it reaches a maximum at the end of 4 or 5 years when its decomposition starts to be equivalent to the annual increments. Losses due to volatilization are of course eliminated in a protected savanna, but we do not known how much of the nitrogen freed during decomposition of the standing biomass is incorporated into the soil and how much is lost in throughfall. It can be assumed, however, that these losses are not very great, because there is little deep drainage and the concentration of minerals in the infiltration water is low.

Another useful comparison is between the nitrogen cycle in a tropical savanna, (burned or protected) and the same cycle in a grassy meadow in a temperate climate. Dahlman and others (1969) worked out a provisional balance of the nitrogen in a prairie ecosystem in Missouri — the first attempt to quantify the nitrogen cycle in a grassland. In comparing their cycle, which is reproduced with slight modifications in figure 38, with ours, several similarities as well as some notable differences are apparent. While the absolute or relative values of nitrogen accumulated in each compartment are quite similar, there is an important difference in two functions of these systems. In the prairie the internal recycling within the vegetable biomass is much less important; moreover, the possible contribution of symbiotic or free nitrogen fixation has not been considered in the prairie system. A much reduced internal recycling in the prairie must mean that the selective pressures during evolution in this direction have been much lower, which can be explained by the much lower risk of losses from external cycling. In fact the prairie climate is much less conducive to losses due to throughfall, as is evidenced by the type of soil evolution and the nutritive status of these soils, both totally different from the profile evolution and extreme nutrient poverty of ferralitic soils. Likewise, the contribution of nitrogen fixation is negligible in a soil rich in nitrogen and therefore unfavorable for nitrogen fixing microorganisms, in addition to the poverty of legumes in this specific prairie. In the tropical savanna the contribution of atmospheric nitrogen through fixation by microorganisms becomes critical in the cycle, since it is this process that contributes most to increase the pool of available nitrogen. The process of free fixation is therefore a key factor in the nutrient cycle of the ecosystems of tropical savannas, as is the process of fixation by symbiotic microorganisms when legumes form part of these ecosystems.

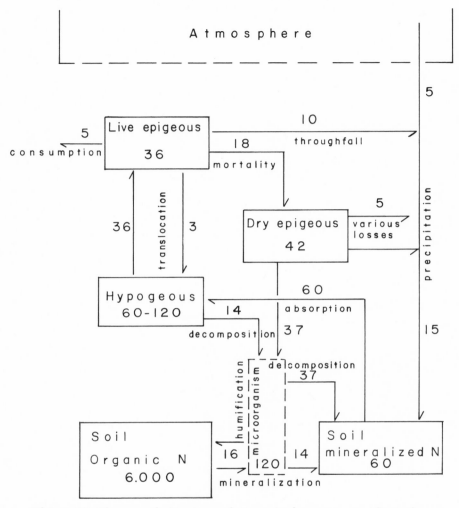

**Figure 38**   The annual nitrogen cycle in an *Andropogon gerardi – Andro-pogon scoparius* prairie in Missouri (U.S.A). Values are in tm.ha$^{-1}$ (stocks in each subsystem) and tm.ha$^{-1}$ · year$^{-1}$ (transfers between subsystems). (Modified from Dahlman et al., 1969.)

## The economy of other nutrients

In the seasonal savannas, phosphorus (fig. 39) found in the soil in absorbable forms (in the horizon where 95% of the roots are found) is available in amounts of 18 kg · ha$^{-1}$. In the vegetation there is at

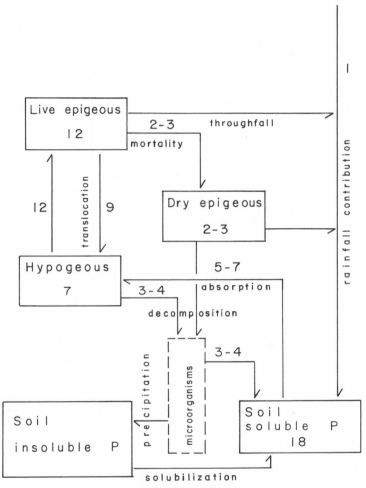

**Figure 39** The annual phosphorus cycle in the seasonal savanna of *Axonopus purpusii – Leptocoryphium lanatum* (Barinas). The values are given in tm.ha$^{-1}$ (stocks in each subsystem) and tm.ha$^{-1}$.year$^{-1}$ (transfers between subsystems).

most a similar amount, which is unevenly divided between the hypogeous and the epigeous biomass in the two seasonal savannas that were studied, in proportion to the respective biomasses. In the savanna of Barinas, with a greater epigeous to hypogeous biomass ratio, 12 kg · ha$^{-1}$ was the maximum amount of phosphorus found in the epigeous portion, and 7 kg $\cdot$ ha$^{-1}$ was found in the hypogeous portion, before the massive translocation that takes place at the end of the growing season, which in equivalent savannas of the llanos has been estimated as 80–90% of the total (Medina et al., 1977). This amount might be an overestimate, since possibly not all the difference between the values of phosphorus in the epigeous biomass obtained at the end of the rainy and dry season may be the result of translocation. Some may have been lost by means of rain throughfall of the decomposing dry biomass during the period between the end of the rainy season and the beginning of the next rainy period. In any case, the internal recycling must consist of more than 50% of the usable phosphorus that circulates through the ecosystem, the other 50% being divided between decomposition of the dead standing biomass and the ephemeral part of the underground organs. While this last can contribute directly to the increase of the pool of usable phosphorus, a part of the nutrient in the dead standing aerial biomass is undoubtedly lost in throughfall.

The nature of the precise relation between total and available phosphorus still remains to be determined. Among the potential reserves are the phosphorus that is immobilized in the humus and the assimilable phosphorus that is outside the reach of the roots of the herbaceous plants. The unavailable phosphorus in primary minerals and other insoluble forms in the soil should also be considered.

The discussion of the cycles of calcium, magnesium, and potassium will be based on the values obtained in the poorer of the two seasonal savannas (Boconoito), where these cations can easily become critical elements. In all three cases the quantity of available cations in the soil is about four times greater than that in the vegetation. Consequently an imbalance can only take place through continued exportation of part of the vegetable biomass, whether through direct harvesting or through consumption by domestic herbivores, without any addition of fertilizers containing these elements. In the hyperseasonal savanna, the cycles of potassium and phosphorus are in a much more labile situation, since most of the available amounts of both elements circulate through the system each year.

In the savanna ecosystems, the soil constitutes the principal reservoir of carbon. The three savannas of the western llanos had soil organic matter contents on the surface of 1.67% (Barinas), 1.78% (Jaboncillo), and 1.89% (Boconoito); the respective quantities of

carbon in the first 100 cm of the soil were 57.7, 40.4, and
98.6 tm · ha$^{-1}$. That is, the pool of carbon in the soil was between 12
and 15 times greater than the amount accumulated in all of the
vegetable biomass at its peak, and much higher than all the carbon
in all the living creatures of the ecosystem, including the fauna and
microorganisms.

# The nutrient balance in tropical rain forests

The first exhaustive investigations of the nutrient economy of a
tropical rain forest were performed by Greenland and Kowal
(1960), Nye and Greenland (1960), and Nye (1961), working in the
semideciduous forest of Kade in Ghana. In addition to these we will
review the results of Golley et al. (1975) from Panama and Hase and
Fölster (1982) in Venezuela, on the American lowland tropical
forest.

For the forest in Ghana, table 28 shows the quantities of five
nutrients in the plant biomass (aerial, underground, and litter), as
well as in the first 30 cm of the soil, where 85.5% of all roots were
found. We have also indicated the respective amounts of the five
nutrients in the annual primary production, which we obtained on
the basis of the production values and of nutrient concentration in
each compartment of the biomass. In this rain forest on poor oxisol,
the recycled quantities of phosphorus, potassium, calcium, and
magnesium in the annual production are between $\frac{1}{2}$ and $\frac{1}{16}$ of the
total available amount of each nutrient in the exchange complex.
The quantities of potassium, calcium, and magnesium accumulated
in the vegetable organic matter are larger than those in the soil pool,
and this is even more notable in the case of phosphorus, twelve
times more of which is retained by the vegetation than is found in
soluble form in the soil. A mean recycling time for each element,
taking into account the quantities accumulated in the vegetation
and the quantities utilized in the annual production, would vary for
the four nutrients between 12 and 16 years.

With regard to nitrogen, table 28 shows that the vegetation and
the soil are equally important, given the order of magnitude of the
amounts retained as a pool of slow recycling. If we now try to
determine the annual cycle of this element in the ecosystem (fig. 40)

**Table 28.** Amounts (in kg/ha) and ratios of mineral elements in the vegetation and the soil of a tropical semideciduous forest in Ghana, on oxisol, and recycling halftimes in vegetation (data from Greenland and Kowal, 1960).

|  | N | P | K | Ca | Mg |
|---|---|---|---|---|---|
| Total biomass: vegetation (V) and litter | 2009 | 137 | 906 | 2670 | 390 |
| Soil (S), in depth to 30 cm | 4592 | 13 | 650 | 2576 | 370 |
| Annual production (P) | 239 | 11 | 74 | 172 | 27 |
| Ratio of nutrients found in P to nutrients found in S, expressed in percentages | 5.2 | 84.6 | 11.4 | 6.7 | 7.3 |
| Ratio of nutrients found in V to nutrients found in V + S, expressed in percentages | 30.4 | 91.3 | 58.2 | 50.9 | 51.3 |
| Recycling halftimes in vegetation, in years | 8 | 12 | 12 | 16 | 14 |

we learn that 1765 kg $\cdot$ ha$^{-1}$ accumulates in the epigeous biomass, of which 1295 kg are in trunks and branches and 470 kg are in branchlets, leaves, flowers, and fruits. Together with the 210 kg $\cdot$ ha$^{-1}$ retained by roots and the 34 kg $\cdot$ ha$^{-1}$ in the litter, a total amount of 2009 kg $\cdot$ ha$^{-1}$ of nitrogen is obtained. We already have pointed out that 4592 kg $\cdot$ ha$^{-1}$ are immobilized in the humus. From the vegetation 199 kg $\cdot$ ha$^{-1}$ of nitrogen pass annually to the litter, to which 36 kg $\cdot$ ha$^{-1}$ must be added as the annual contribution from rotting wood to this compartment. Rainfall washes a small proportion of nitrogen from leaves and litter, which, together with the rainfall contribution, represents an annual gain of 27 kg $\cdot$ ha$^{-1}$ for the mineral nitrogen in the soil. The amount of nitrogen utilized in the annual increase of the roots was estimated at 30 kg $\cdot$ ha$^{-1}$, which must be matched by the amount decomposed each year if this nitrogen compartment is to remain constant for an extended time period.

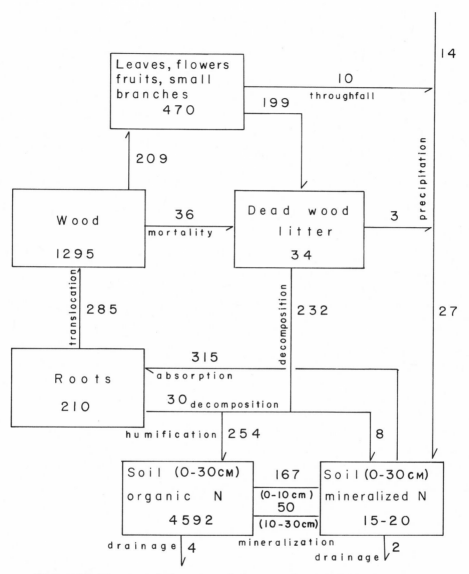

**Figure 40** The annual nitrogen cycle in a semideciduous forest in Ghana. Values are in tm.ha$^{-1}$ (stocks in each subsystem) and tm.ha$^{-1}$·year$^{-1}$ (transfers between subsystems). (Data taken from Greenland and Kowal, 1960, Nye and Greenland, 1960, and Nye, 1961.)

The quantity of organic nitrogen mineralized each year in the soil and in the litter by the action of microorganisms was not determined in these studies on the tropical rain forest of Kade in Ghana. But working in similar conditions on the Ivory Coast, Rham (1969) obtained an annual value of nitrogen through mineralization of 140 to 190 kg $\cdot$ ha$^{-1}$. In the same forest Bernhard-Reversat (1974) determined a microbial production of mineral nitrogen of 167 kg $\cdot$ ha$^{-1}$ in the 0–10 cm horizon, plus 8 kg $\cdot$ ha$^{-1}$ $\cdot$ year$^{-1}$ produced by the mineralization of the nitrogen in the litter. The mineralization of deeper horizons, up to 30 cm, represents approximately 30% of that produced between 0 and 10 cm, or about 50 additional kg $\cdot$ ha$^{-1}$. Applying these values to the forest of Ghana yields the following calculation: 217 kg $\cdot$ ha$^{-1}$ are mineralized annually from the organic nitrogen in the humus and 8 kg $\cdot$ ha$^{-1}$ from the litter (since the total contribution of the vegetation including roots, is 262 kg $\cdot$ ha$^{-1}$), so that 254 kg $\cdot$ ha$^{-1}$ would be incorporated into the soil through humification, which implies that 37 kg $\cdot$ ha$^{-1}$ of nitrogen are incorporated into this compartment above the balance between input and output. Even assuming that a certain amount is lost by infiltration, these values represent an annual increase of 0.7% of the organic nitrogen in the soil. On the basis of the amount of water infiltrated into these forests (Huttel, 1975), estimated in the order of 600 mm $\cdot$ year$^{-1}$ and of the concentration of 1.06 ppm nitrogen in the drainage water (Bernhard-Reversat, 1975), the losses are 60% as organic nitrogen and 40% as mineral nitrogen.

The losses of nitrogen from the vegetation plus the net annual accumulation must of course equal the total annual absorption. The net annual accumulation is 40 kg $\cdot$ ha$^{-1}$ (Nye, 1961) and the losses are 199 kg $\cdot$ ha$^{-1}$ as litter, 10 kg $\cdot$ ha$^{-1}$ as throughfall, 36 kg $\cdot$ ha$^{-1}$ by decomposition of dead wood, and 30 kg $\cdot$ ha$^{-1}$ from hypogeous decomposition, for a total of 315 kg $\cdot$ ha$^{-1}$. Consequently an equal amount must move annually from the soil to the vegetation. If mineralization provides 217 kg $\cdot$ ha$^{-1}$, mineralization of litter 8 kg $\cdot$ ha$^{-1}$ and throughfall contributes 27 kg $\cdot$ ha$^{-1}$, the total income to the soil mineral nitrogen pool is 252 kg $\cdot$ ha$^{-1}$. The shortage of 65 kg $\cdot$ ha$^{-1}$, which represents the excess of losses over income in this compartment of the system, requires an explanation.

The missing amount may be produced through nitrogen fixation by free or symbiotic microorganisms, although Bernhard-Reversat estimates that the addition of nitrogen through this mechanism is

rather low. Some nitrates are probably lost through bacterial denitrification, but since this only takes place under infrequent anaerobic conditions when the soil is waterlogged, the author assumes
that this process has no quantitative importance within the yearly
balance.

The nutrient economy of the semideciduous forest in the proximity of Santa Fe (Panama), which has many floristic and ecological
similarities with the semideciduous forests of the western llanos of
Venezuela (Sarmiento et al., 1971) and of the llanos of Colombia
(FAO, 1966), has been summarized in table 25, which provides the
pertinent data for four elements: phosphorus, potassium, calcium,
and magnesium. Again it is evident that the principal pool of
nitrogen and potassium is the vegetation, where close to 90% of the
total amounts of these elements in the ecosystem accumulate. On
the other hand, 4 to 5 times more calcium and magnesium are
found in the soil than in the vegetation. This difference with the
African forest is probably the result of the evolution of the soil in the
Panamanian forest, which has developed over a substrate very rich
in these two elements. Moreover, the results that we reproduce
from Golley et al. (1975) refer only to the total amounts of these
metals in the soil, not the quantity that is soluble or utilizable by the
vegetation.

In a summary of their research, Golley and his colleagues group
the mineral elements in four categories, according to their recycling time and to the amounts present in the soil, as follows:

1. An element that is present only in small quantities in the soil
and also has a fast recycling rate. Phosphorus is the only element in
this category, and according to these authors could be the most
limiting nutrient in this ecosystem.

2. Elements that exist in the soil in small amounts but have a slow
recycling rate. In this category are included all the elements that do
not belong in the other three categories.

3. Elements that have an important soil pool and a fast cycle. The
only nutrient with this characteristic is potassium, which consequently also could be limiting.

4. Elements with large amounts in the soil pool and with slow
recycling rates. To this category belong calcium, magnesium, and
sodium.

The results of Hase and Fölster (1982) for the forest on alluvial soils in the western Venezuelan llanos, summarized in table 26, lead to essentially the same conclusions. The bulk of the nitrogen, phosphorus, and magnesium is stored in the soil, whereas most of the calcium and potassium is stored in the vegetation.

Potassium and phosphorus, even though not limiting under present conditions in these forests, could conceivably become so; both nutrients, however, are recycled within the ecosystem (between the vegetation and the soil) with cycles that are 17 to 21 times greater than what it takes to pass from the soil to the drainage water. Calcium might also be limiting when the soil is not particularly rich in this ion, which is the case of the forests of the western llanos.

In conclusion, the most important difference in the nutrient economy of tropical forests and savannas is in the location of the principal pool of an element in the ecosystem. In the forests the quantities of potassium and phosphorus accumulated in the vegetation are larger than those in the soil, both their soluble forms and total amounts. Nitrogen, on the other hand, has approximately the same importance in the two compartments, while calcium and magnesium may accumulate mostly in the soil or in the vegetation, depending on the type of soil.

The savannas present a totally opposite picture. The pool of soil nitrogen is many times greater than the amount accumulated in the plant biomass; and so also are the soil quantities of calcium, magnesium, and potassium, which are greater than those in the vegetation at any time in the annual cycle. Only the phosphorus is equally divided between the two fundamental compartments of the ecosystem.

These results point out that the forests represent a significant accumulation of mineral nutrient, so that the disappearance of the vegetation would leave the system notably impoverished. In other words, the fundamental strategy of the trees in these ecosystems is to accumulate and concentrate an appreciable amount of nutrients. In the savannas, on the other hand, the quantities concentrated in the vegetation, with the notable exception of phosphorus, are but a fraction of the quantities of assimilable nutrients available in the soil. But because these quantities are very low to begin with, the systematic elimination of the vegetation through any management system would soon lead to a system exceptionally deficient in mineral nutrients.

# 7

# Synthesis and conclusions

A final evaluation of the research results leads to the following generalizations:

1. Despite their great diversity, neotropical savannas conform to an ecological-physiognomic type of natural ecosystem that is found exclusively in warm and humid tropical areas (the low-altitude wet tropics). Their essential characteristic is a herbaceous cover of perennial grasses and sedges, with the typical "bunch grass" type of growth; very often they also support an open tree stratum. Fire has been a normal ecological factor in their evolution and the selective pressure of fire has been a crucial factor in the evolution of many morphofunctional aspects of savanna species, both woody and herbaceous.

2. All American savannas are intrinsically seasonal systems, in the sense that their principal functional processes proceed at very different rates in different seasons of the year. There is always a period of reduced activity of the herbaceous cover, coinciding with an edaphic drought season of varying length that affects the components of the herbaceous layer. This xeropause, common to all savanna systems, may be accompanied by the opposite stress produced by an excess of water in the soil. According to the intensity of these opposite kinds of stresses we have distinguished four ecological savanna types, which we have called semiseasonal, seasonal, hyperseasonal, and marshy savannas.

3. The herbaceous biomass of the savannas has a continuous vertical distribution both above and below the surface; the same is

true for the aerial biomass of the woody stratum. The greatest accumulation of living and dead biomass takes place in the vegetation layers closest to the surface, so that more than two-thirds of the biomass is generally found between 20 cm below and above the ground. The herbaceous stratum of American savannas reaches maximum heights of a little over 80 cm, while the grass inflorescences reach up to 140 cm. The only exceptions are found in some flooded savannas, which have taller vegetation. The tallest trees rarely surpass 8 m; the greatest foliage biomass is found mostly between 2 and 4 m. Nor do these trees form thick trunks, from which it may be assumed that they do not reach the ages of forest trees.

4. The hypogeous biomass diminishes very rapidly with depth, and it is rare to find more than 10% of this biomass below 100 cm, while below 200 cm roots are almost never found. Only tree roots can reach lower levels (down to 16 m), especially in well-drained soils. The concentration of the greatest proportion of the underground biomass in the upper centimeters of the soil is more accentuated in hyperseasonal savannas, where the water table may reach one or two meters but cannot be utilized by the herbaceous species because they cannot grow through the hard clay horizons and must limit themselves to exploiting the water available in superficial horizons above the clay pan.

5. Two elements form the horizontal structure of the herbaceous cover: the space occupied by the perennial grasses and sedges, and the free interspaces. This differentiation can become accentuated by later processes of a physical (sheet erosion) or biological (worm casts) nature. Trees also form a structural element that creates horizontal microhabitat differences and leads to the creation of new niches. According to the density and cover of the woody stratum, savannas can be physiognomically classified as: grassy savannas, open savannas, closed savannas, and woodland, but even in this last case, when the tree cover reaches 30–40%, the principal flow of organic matter, energy, water, and nutrients is through the herbaceous stratum. Consequently we maintain that this is the ecologically dominant and characteristic element of the savanna ecosystem.

6. The most interesting morphological adaptation of savanna species—one that portrays the stress to which they are subjected —is the burying of the biomass, which is evident not only in the

species of the herbaceous stratum, but also, and at times in a very spectacular way, in the form of enlarged underground organs in the woody species, both shrubs and trees, and occasionally even palms.

7. The characteristics of the physical environment, as well as the variations induced by the extraordinary structural and functional changes of the savanna ecosystem throughout the annual cycle, have led to the evolution of a diversity of phenological strategies that conform to the spatial and temporal availability of resources. Certain strategies, such as those of the precocious species, for example, take advantage of resources not used during certain times of the year by the dominant species, thereby saturating the phenological spectrum of these ecosystems.

8. No fundamental process of the savanna ecosystem can be understood if it is not considered in the light of the intrinsic seasonality of these ecosystems. For example, the basic processes that constitute the productive process—assimilation, mortality, reallocation, decomposition—take place more or less without interruption, but at very different rates in each phenophase of each of the basic phenological strategies. Consequently, the global rates of each process in the annual cycle will depend on the complexity and integration of the various constituent strategies. It is very difficult to obtain a general understanding of the productive process from simple parameters like the peak of maximum green biomass. Instead, one must determine the productive rates of each group of species. Since these studies are still in their early stage, several essential aspects of the productive processes of tropical savannas remain unknown, especially the interrelationship between the undergound and aerial subsystems and its effects on the respective production levels, as well as the seasonal variations in the production, mortality, and decomposition of each of these compartments within the total productive system of the savanna.

9. Because the evaluation of productive processes lacked the necessary precision, it is certain that the published values for savanna productivity underestimate the real values. It is likely that they are 50 to 100% below the real values. The divergence will be greater the more complex and structured a savanna is, the greater the diversity of species' strategies, and the more evenly represented the various species are within the dominant flora of the ecosystem.

10. The species with successful strategies are the ones that optimize the carbon balance during a relatively extensive period and during the rest of the cycle survive the constraints of the environment, whether physical stress or biological competition, by means of a "defensive" strategy. Such strategies allow the species to survive and complete their life cycle but result in low levels of photosynthetic efficiency for the totality of the ecosystem. Instead, species succeed each other in the utilization of limited resources in each vegetation cycle, each coming forward to exploit the system for a short period only to be replaced by another. It is not useful to qualify these strategies as "stress resistant," "competitive," or "disturbance" type (in the sense used by Grime, 1977), since all these terms are applicable. Furthermore, a group of species may exhibit the characteristics of one type at one time, and of another some time later.

11. A good example of the spatial and temporal partition of resources is the manner of utilization of water, which the vegetation exploits as a limiting and scarce resource. This becomes evident in a comparison of the behavior of herbaceous and woody species, but is also present among the different strategies found in the herbaceous stratum.

12. When the yearly water balance of savannas and tropical forests is compared, the greatest difference is not in the quantity of water lost through evapotranspiration, but in the amount that filters down toward the lower soil levels. In the forests, infiltration represents a substantial part of the yearly water balance, while in savannas the balance between pluvial inputs and evapotranspiration losses is so tight, that only in exceptionally wet years is there an excess of water that is lost through surface runoff or through deep infiltration. The implications of this behavior for the management of watersheds and the rational use of water requires no comment.

13. With few exceptions, the economy of several essential nutrients constitutes one of the most labile aspects of savanna ecosystems, indeed, of most tropical ecosystems. However, the adaptive mechanisms of savannas and forests diverge in confronting a dystrophic environment. Forests may occupy relatively rich soils as well as some very poor ones, and consequently offer a greater gamut of adaptations than do savannas, which occupy only poor soils. The adaptive solutions of the forests are based on the existence of a nutrient pool within the green biomass itself, as well as on

short-circuiting to a large extent the normal cycling of nutrients through the soil by enclosing them within the vegetation. But savannas occupy the poorer sites, and the fast mineral cycling rates exclude the possibility of maintaining large reserves within the biomass, and the principal mineral reservoir therefore becomes the soil.

14. The nutrient poverty of savanna soils is an environmental factor largely independent of the vegetation, since it is related in great measure to geomorphogenetic and pedogenetic processes that take place at the level of the landscape and natural formation. However, once oligotrophism displaces the forest/savanna equilibrium and the area turns into savanna, its specific characteristics tend to reinforce the impoverishment of the system, either through periodical fires, or because of the low biomass stock.

15. The cycling of nitrogen is one of the most vulnerable aspects of the savanna ecosystem. The major reservoir of nitrogen appears to be the organic matter in the soil, in the form of stable humic compounds, of high molecular weight, which are biodegraded very slowly. This pool of slow recycling acts as a nitrogen-conserving, homeostatic element. The amount of nitrogen that circulates through the system each year, whether in the form of ammonium or of nitrates, is almost half the amount of the organic nitrogen, so that there is never any nitrogen that is not immediately incorporated into the biomass of producers or decomposers. It must also be pointed out that all the processes of nitrogen transfer are mediated by specific microorganisms that insure the decomposition of vegetable matter, the mineralization, humification, free and symbiotic fixation, denitrification, and so on. These biochemical and microbiological processes are also essentially labile and depend for their development on well-defined microenvironmental conditions that do not tolerate major disruptions.

16. Potassium, phosphorus, and calcium are resources in short supply that must be used carefully to insure the continued functioning of the savanna ecosystem. As in the case of nitrogen, these elements are recycled within the vegetation, thereby reducing the losses due to volatilization or lixiviation and insuring that they may be reused by the plant when needed. The great importance of this recycling of nutrients makes it urgent to reach a better understanding of the dynamic behavior of the hypogeous biomass, of the rates of recycling of these elements in different organs (roots,

rhizomes, xylopods, and so on), and of the changing metabolic interrelationships between the underground organs and the assimilatory and reproductive organs.

## Elements for an interpretation of savanna ecosystems

The most significant relationships between the different components of the ecosystem and the environmental factors that regulate the external stimuli are graphically summarized in figures 41, 42, and 43.

We have outlined the external factors whose effects cannot be biologically controlled by the ecosystem; these constitute the independent variables, or primary elements of the system (fig. 41). For

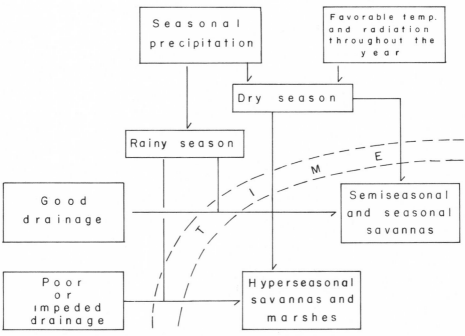

**Figure 41**  A schematic representation of the impact of primary environmental factors on the different types of tropical savanna ecosystems, determining over time a series of consequences and ecosystem responses described in figures 42 and 43.

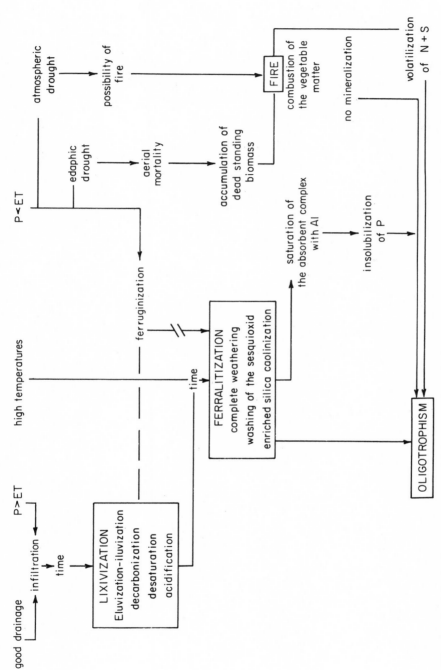

**Figure 42** Principal consequences of the action of the primary factors on the ecosystem of the tropical seasonal savanna.

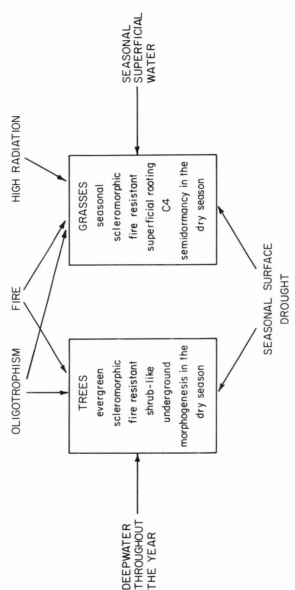

**Figure 43** Principal responses of the dominant life forms in the ecosystem of the tropical seasonal savanna to the primary environmental factors and their ecological consequences.

all the tropical savannas these primary elements are: high and low rainfall seasons; high temperatures and high solar radiation throughout the year; and a long period of exposure to each of these elements. Drainage is another differentiating element of the various savanna types. If drainage is adequate, conditions develop that lead to seasonal or semiseasonal savannas, while insufficient drainage produces conditions that result in hyperseasonal savannas or marshes.

To be more precise, the period of high rainfall is known to last six to nine months, each month producing over 100 mm — sometimes considerably more. The period of low rainfall may be as long as six months or as short as one or two, with the rest of the season being intermediate in nature. During the dry months there is either no rainfall or less than 50 mm per month. By constant high temperatures we mean mean daily temperatures that are always above 18° C; there are no low temperatures to limit physiological processes. Solar radiation will have the characteristic pattern of low latitudes with little variation throughout the year. A long period of evolution under these conditions means a temporal scale that is measured in units of 10,000 to 100,000 years at least.

These conditions are found in the humid tropics under a seasonal climate that normally coincides with the *Aw* and *Am* types of the Koeppen system. However, if the rainfall during the wet season is very high and the dry season is very short, the conditions tend toward an ever-wet climate of *Af* type. On the other hand, the temporal scale leads us into relatively old areas, where weathering and pedogenesis have reached the limit of their effects within the specified morphoclimate.

With this set of concurrent primary factors, and using as example a site with well-drained soils, figure 42 depicts the principal processes of ecological importance that will lead to certain effects, among them an increasing nutrient deficiency. These effects are amplified and reinforced by the biotic responses of the ecosystem. Finally, figure 43 describes a series of adaptive responses of a morphological or functional nature that have evolved as adaptive survival strategic responses of the species to the conditions that have developed. We have also indicated some elements of the interactions between producers and consumers that may reinforce the primary responses of the former.

The three diagrams together constitute an integral picture of the principal lines of reasoning that must be followed to unravel the evolutionary complexity inherent in tropical savanna ecosystems.

# References

Ab'Saber, A. 1963. Contribuição a geomorfología da área dos cerrados. In *Simpósio sobre o Cerrado*, pp. 119–124. São Paulo, Editóra da Universidade de São Paulo.

——— 1973. A organização natural das paisagens inter e subtropicais brasileiras. In *III Simpósio sobre o Cerrado*, pp. 1–14. São Paulo, Editóra da Universidade de São Paulo.

Ahmad, N., and R. L. Jones. 1969. A plinthaquult of the Aripo savannas, North Trinidad. I. Properties of the soil and chemical composition of the natural vegetation. *Soil Sc. Soc. Amer. Proc.* 33:762–768.

Alexander, E. B. 1973. A comparison of forest and savanna soils in northeastern Nicaragua. *Turrialba* 23:181–191.

Alvim, P. de T., and W. Araujo. 1952. El suelo como factor ecológico en el desarrollo de la vegetación en el centro-oeste del Brasil. *Turrialba* 2:153–160.

Ambasht, R. S., A. N. Maurya, and U. N. Singh. 1972. Primary production and turnover in certain protected grassland of Varanasi, India. In P. M. Golley and F. B. Golley, eds., *Tropical Ecology*, pp. 43–50. Athens, Ga, University of Georgia Press.

Anderson, A. B. 1981. White-sand vegetation of Brazilian Amazonia. *Biotropica* 13:199–210.

Arens, K. 1958. O cerrado como vegetação oligotrófica. *Bol. Fac. Fil. Cienc. Letr. USP* 224, *Botanica* 15:58–77.

——— 1963. As plantas lenhosas dos campos cerrados como flora adaptada as deficiências minerais do solo. In *Simpósio sobre o Cerrado*, pp. 287–299. São Paulo, Editóra da Universidade de São Paulo.

Aristeguieta, L. 1966. Flórula de la estación biológica de Los Llanos. *Bol. Soc. Venez. Cienc. Nat.* 110:228–307.

Askew, G. P., D. J. Moffat, R. F. Montgomery, and P. L. Searl. 1971. Soils and soil moisture as factors influencing the distribution of the vegetation formations of the Serra do Roncador, Mato Grosso. In *III Simpósio*

## 218   References

*sobre o Cerrado,* pp. 150–160. São Paulo, Editóra da Universidade de São Paulo.

Ataroff, M. 1976. Estudios ecológico-poblacionales en dos especies de árboles de las sabanas de los Llanos. Ph.D diss., University of Mérida.

Balandreau, J. 1976. Fixation rhizosphérique de l'azote en savane de Lamto. *Rev. Ecol. Biol. Sol.* 13:529–544.

Barrios, S., and V. González. 1971. Rhizobial symbiosis on Venezuela savannas. *Plant & Soil* 34:707–719.

Bartholomew, W. V., J. Meyer, and H. Landelout. 1953. Mineral nutrient immobilisation under forest and grass fallow in the Yangambi (Belgian Congo) region, with some preliminary results on the decomposition of plant material on the forest floor. *Publ. Inst. Nat. Etude Agron. Congo Belge, Ser. Sci.* 57:1–27.

Bazilevich, N. I., and L. E. Rodin. 1966. The biological cycle of nitrogen and ash elements in plant communities of the tropical and subtropical zones. *Forestry Abstr.* 27:357–368.

Beard, J. S. 1953. The savanna vegetation of northern tropical America. *Ecological Monographs* 23:149–215.

Bernhard-Reversat, F. 1974. L'azote du sol et sa participation au cycle biogéochimique en forêt ombrophile de Côte d'Ivoire. *Rev. Ecol. Biol. Sol.* 11:263–282.

——— 1975. Recherches sur l'ecosystème de la forêt subéquatoriale de basse Côte-d'Ivoire. VI. Les cycles des macroélements. *Terre Vie* 29:229–254.

——— 1982. Biogeochemical cycle of nitrogen in a semi-arid savanna. *Oikos* 38:321–332.

Blanck, J. P. 1976. Repartition de la savane et de la forêt dans les Llanos Centrales du Vénézuela (Vallée du Rio Orituco). *Photointerpretation* 76:1–10.

Blanck, J. P., L. Vivas, F. Salas, J. B. Castillo, M. Trucic, R. Marante, and O. Cabello. 1970. *Estudio de los suelos del área del Ticoporo I (Llanos Altos Occidentales de Barinas).* Mérida, Instituto de Geografía, Fac. de Ciencias Forestales, U.L.A.

Blanck, J. P., M. Monasterio, and G. Sarmiento. 1972. Las comunidades vegetales y su relación con la evolución cuaternaria del relieve en los Llanos Centrales de Venezuela. *Acta Cient. Venez.* 23(supl. 1):31.

Blydenstein, J. 1962. La sabana de *Trachypogon* del alto llano. *Bol. Soc. Venez. Cienc. Nat.* 102:139–206.

——— 1963. Cambios en la vegetación después de protección contra el fuego. I. El aumento anual en material vegetal en varios sitios quemados y no quemados en la estación biológica. *Bol. Soc. Venez. Cienc. Nat.* 103:233–38.

Bourlière, F., ed. 1983. *Tropical Savannas.* Amsterdam, Elsevier.

Bourlière, F., and M. Hadley. 1970. The ecology of tropical savannas. *Ann. Rev. Ecol. Syst.* 1:125–152.

Bulla, L. A., J. Pacheco, and R. Miranda. 1977. Producción, descomposición y dinámica de una sabana bajo diferentes condiciones de inundación. *IV Simp. Int. Ecol. Trop. Panamá, Resúmenes:*22–24.

———— 1980a. Producción, descomposición, flujo de materia orgánica y diversidad en una sabana de banco del Módulo Experimental de Mantecal (Estado Apure, Venezuela). *Acta Cient. Venez.* 31:331–338.

———— 1980b. Ciclo estacional de la biomasa verde, muerta, y raíces en una sabana inundada de estero en Mantecal (Venezuela). *Acta Cient. Venez.* 31:339–344.

Buringh, P. 1970. *Introduction to the study of soils in tropical and subtropical regions.* Wageningen, Centre for Agricultural Publishing and Documentation.

Cabrera, A. L., and A. Willink. 1973. *Biogeografía de América Latina.* Washington, D.C., OEA, Monografías Científicas, Serie Biológica, 13.

Canales, M. J., and J. F. Silva. 1982. Efecto del fuego sobre el crecimiento vegetativo y reproductivo de *Sporobolus cubensis,* gramínea precoz de la sabana. *Acta Cient. Venez.* 33 (suppl. 1):108.

Castillo, J. B., L. Vivas, M. Trucic, R. Marante, and J. P. Blanck. 1972. *Estudio de los suelos de las áreas de Ticoporo II–III. Llanos Altos Occidentales de Barinas.* Mérida, Instituto de Geografía y Conservación de Recursos Naturales, Fac. Ciencias Forestales, U.L.A.

Cesar, J. 1971. Etude quantitative de la strate herbacée de la savane de Lamto (Moyenne Côte d'Ivoire). PhD. diss., Faculté des Sciences de Paris.

Cesar, J., and J. C. Menaut. 1974. Le peuplement végétal des savanes de Lamto (Côte d'Ivoire). *Bull. Liaison cherch. Lamto, Numéro spécial 1974, Fasc. II*:1–161.

Clark, F. E., and E. A. Paul. 1970. The microflora of grassland. *Adv. Agronomy* 22:375–435.

Clarkson, D. T. 1969. Metabolic aspects of aluminium toxicity and some possible mechanisms for resistance. In I. H. Rorison, ed., *Ecological Aspects of The Mineral Nutrition of Plants,* pp. 381–397. Oxford, Blackwell.

Clements, F. E. 1916. *Plant Succession.* Washington, D.C., Carnegie Inst. Wash., Publ. 242.

Coleman, N. T., J. T. Thorup, and W. A. Jackson. 1960. Phosphate-sorption reactions that involve exchangeable Al. *Soil Sc.* 90:1–7.

Comerma, J. A., and O. Luque. 1971. Los principales suelos y paisajes del Estado Apure. *Agron. Trop.* 21:379–396.

Coutinho, L. M., and A. Lamberti. 1971. Algumas informaçoes sobre a análise de solos sob Mata de Terra Firme e Mata de Igapó. *Ciencia e Cultura* 23:601–603.

Dahlman, R. C., and C. L. Kucera. 1965. Root productivity and turnover in native prairie. *Ecology* 46:84–89.

Dahlman, R. C., J. S. Olson, and K. Doxtader. 1969. The nitrogen economy of grassland and dune soils. In *Biology and Ecology of Nitrogen*, pp. 54–89. Washington, D.C., National Academy of Sciences.

Dansereau, P. 1957. *Biogeography, an Ecological Perspective.* New York, Ronald Press.

Daubenmire, R. 1972. Standing crop and primary production in savanna derived from semideciduous forest in Costa Rica. *Bot. Gaz.* 133:395–401.

Dirven, J. 1963. The nutritive value of the indigenous grasses of Surinam. *Neth. J. Agric. Sci.* 11:295–307.

Dobereiner, J. 1979. Fixação de Nitrogênio em gramíneas tropicais. *Interciencia* 4:200–206.

Dobereiner, J., and J. M. Day. 1974. Importancia potencial de la fijación simbiótica de nitrógeno en la rizósfera de gramíneas tropicales. In E. Bornemisza and A. Alvarado, eds., *Manejo de Suelos en la América Tropical*, pp. 203–216. Raleigh, N.C., Soil Science Dept., North Carolina State University Press.

Eden, M. J. 1964. *The Savanna Ecosystem-Northern Rupununi, British Guiana.* McGill University Savanna Research Series 1. Montreal, McGill University Press.

——— 1974. Paleoclimatic influences and the development of savanna in southern Venezuela. *J. Biogeogr.* 1:95–109.

Egler, W. A. 1960. Contribuiçoes ao conhecimento dos campos da Amazonia. I. Os campos do Ariramba. *Bol. Museu Paraense Emilio Goeldi, Bot.* 4:1–36.

Egunjobi, K. K. 1974. Dry matter, nitrogen, and mineral element distribution in an unburnt savanna during the year. *Oecol. Plant.* 9:1–10.

Eiten, G. 1972. The cerrado vegetation of Brazil. *Bot. Rev.* 38:201–341.

——— 1977. Delimitação do conceito de cerrado. *Arq. Jardim Bot.* 21:125–135.

Ellemberg, H. 1971. Nitrogen content, mineralization, and cycling. In Duvigneaud, P., ed., *Productivity of Forest Ecosystems*, pp. 509–514. Paris, UNESCO.

Entrena, J., V. Gonzalez, and L. Bulla. 1977. Variaciones en la composición florística, biomasa y producción neta en una sabana del Sur de Venezuela. In *IV Simp. Int. Ecol. Trop., Panamá, Resúmenes*, pp. 34–35. Panama, Int. Soc. of Tropical Ecology.

Ernst, W. 1975. Variation in the mineral contents of leaves of trees in miombo woodlands in South Central Africa. *J. Ecol.* 63:801–807.

Escobar, A. 1977. Estudio de las sabanas inundables de *Paspalum fasciculatum*. PhD. diss., Instituto Venezolano de Investigaciones Científicas.

Escobar, A., and E. González Jiménez. 1975. Production primaire de la savane inondable. In *III Simp. Inter. Ecol. Trop., Lubumbashi.* Lubumbashi, Int. Soc. of Tropical Ecology.

——— 1979. La production primaire de la sabane inondable d'Apure (Venezuela). *Geo-Eco-Trop.* 3:53–70.

Escobar, A., and E. Medina. 1977. Estudio de las sabanas inundables de *Paspalum fasciculatum.* In *IV Simp. Int. Ecol. Trop.,* Panamá, Resúmenes, pp. 36–37. Panama, Int. Soc. of Tropical Ecology.

Espinoza, D. C. 1969. Variación estacional de los constituyentes bromatológicos de la paja peluda (*Trachypogon plumosus* Humb. et Bonp. Nees) en tres zonas de sabana. *Acta Bot. Venez.* 4:389–423.

F.A.O. 1964. *Reconocimiento edafológico de los Llanos Orientales, Colombia. I. Informe general.* Rome, FAO.

——— 1965. *Reconocimiento edafológico de los Llanos Orientales, Colombia. II. Los suelos de los Llanos Orientales.* Rome, FAO.

——— 1966. *Reconocimiento edafológico de los Llanos Orientales, Colombia. III. La vegetación natural y la ganadería en los Llanos Orientales.* Rome, FAO.

Ferri, M. G. 1944. Transpiração de plantas permanentes dos "cerrados." *Bol. Fac. Fil. Cienc. Letr. U.S.P.* 41, *Botanica* 4:159–224.

——— 1955. Contribuição ao conhecimento da ecologia do cerrado e da caatinga. Estudo comparativo do balanco d'agua de sua vegetação. *Bol. Fac. Fil. Cienc. Letr. U.S.P.* 195, *Botanica* 12:1–170.

——— 1963. Histórico dos trabalhos botánicos sobre o cerrado. En *Simpósio sobre o Cerrado:*19–50. São Paulo, Editóra da Universidade de São Paulo.

Ferri, M. G., and L. M. Coutinho. 1958. Contribuição ao conhecimento da ecología do cerrado. Estudo comparativo da economía d'agua da sua vegetação em Emas (São Paulo) e Goiania (Goias). *Bol. Fac. Fil. Cienc. Letr. U.S.P.* 224, *Botanica* 15:103–150.

Foldats, E., and E. Rutkis. 1965. Influencia mecánica del suelo sobre la fisionomía de algunas sabanas del llano venezolano. *Bol. Soc. Ven. Cienc. Nat.* 108:335–392.

——— 1969. Suelo y agua como factores determinantes en la selección de algunas de las especies de árboles que en forma aislada acompañan nuestros pastizales. *Bol. Soc. Ven. Cienc. Nat.* 115–116:9–30.

——— 1975. Ecological studies of chaparro (*Curatella americana* L.) and manteco (*Byrsonima crassifolia* H.B.K.) in Venezuela. *J. Biogeogr.* 2:159–178.

French, M. H., and L. M. Chaparro. 1960. Contribución al estudio de la composición química de los pastos en Venezuela durante la estación seca. *Agron. Trop.* 10:57–69.

Gadgil, M., and O. Solbrig. 1972. The concept of r- and K- selection: evidence from wild flowers and some theoretical considerations. *Am. Natur.* 106:14–31.

Galrao, E. Z., and A. S. Lopes. 1980. Deficiências nutricionais em solos de cerrado. In *Simpósio sobre o Cerrado V. Cerrado: uso e manejo.* Brasilia. Brasilia, Ed. Editerra.

Godron, M. 1968. *Code pour le relevé méthodique de la végétation et du milieu.* Paris, C.N.R.S.

Goldstein, G., G. Sarmiento, and F. Meinzer. 1982. Un análisis de la economía hídrica en especies de la sabana estacional y su interpretación ecológica y fisiológica. *Acta Cient. Venez.* 33 (suppl. 1):114.

Golley, F. B. 1965. Structure and function of an old field broomsedge community. *Ecological Monographs* 35:113–131.

Golley, F. B., J. T. McGinnis, R. G. Clements, G. I. Child, and M. J. Duever. 1975. *Mineral Cycling in a Tropical Moist Forest Ecosystem.* Athens, Ga., University of Georgia Press.

Gomide, J. A., C. H. Noller, G. O. Mott, J. H. Conrad, and D. L. Hill. 1969. Mineral composition of six tropical grasses as influenced by plant age and nitrogen fertilization. *Agron. J.* 61:116–120.

González, J. A. 1977. Valor nutritivo y contenido mineral en relación a parámetros ambientales y la fenología de dos pastos tropicales: *Leersia hexandra* (Swartz) y *Axonopus purpusii* (Mez) Chase. Caracas, Facultad de Ciencias, U.C.V.

González Jiménez, E. 1979. Primary and secondary productivity in flooded savannas. In *Tropical Grazing Land Ecosystems: A State-of-knowledge Report Prepared by UNESCO/UNEP/FAO*, pp. 620–625. Paris, UNESCO.

González Jiménez, E. and A. Escobar. 1975. Superposition des regimes alimentaires de quatre herbivores de la savane inondable. In *III Simp. Int. Ecol. Trop. Lubumbashi.* Lubumbashi, Int. Soc. of Tropical Ecology.

—— 1977. Adaptación a las condiciones de inundación, productividad y valor nutritivo de las gramíneas de la sabana inundable. In *IV Simp. Int. Ecol. Trop., Panamá, Resúmenes*, pp. 43–44. Panama, Int. Soc. of Tropical Ecology.

Goodland, R. 1971a. A physiognomic analysis of the "cerrado" vegetation of Central Brazil. *J. Ecol.* 59:411–419.

—— 1971b. Oligotrofismo e aluminio no cerrado. In *III Simpósio sobre o Cerrado*, pp. 44–60. São Paulo, Editóra da Universidade de São Paulo.

Goodland, R., and R. Pollard. 1973. The Brazilian cerrado vegetation: a fertility gradient. *J. Ecol.* 61:219–224.

Goodland, R., and M. G. Ferri. 1979. *Ecologia do Cerrado.* Belo Horizonte, Livraria Itatiaia Editora.

Gottlieb, O. R., J. de O. Meditsch, and M. T. Magalhaes. 1966. Com vistas ao aproveitamento do cerrado como ambiente natural. Composição química de especies arbóreas. In L. G. Labouriau, ed., *II Simpósio sobre o Cerrado*, pp. 303–314. *Anais Acad. Bras. Cienc.* 38 (supl).

Greenland, D. J. and J. M. Kowal. 1960. Nutrient content of the moist tropical forest of Ghana. *Plant & Soil* 12:154–174.

Greig-Smith, P. 1964. *Quantitative Plant Ecology.* 2nd ed. London, Butterworth.

Grime, J. P., and J. Hodgson. 1969. An investigation of the ecological significance of lime-chlorosis by means of large-scale comparative experiments. In I. Rorison, ed., *Ecological Aspects of the Mineral Nutrition of Plants*, pp. 67–99. Oxford, Blackwell.

Grime, J. P. 1974. Vegetation classification by reference to strategies. *Nature* 250:26–31.

—— 1977. Evidence for the existence of three primary strategies in plants and its relevance to ecological and evolutionary theory. *Am. Natur.* 111:1169–1194.

Hase, H., and H. Fölster. 1982. Bioelement inventory of a tropical (semi-) evergreen seasonal forest on eutrophic alluvial soils, Western Llanos, Venezuela. *Acta Oecologica, Oecol. Plant.* 3:331–346.

Herrera, R., C. F. Jordan, H. Klinge, and E. Medina. 1978. Amazon ecosystems. Their structure and functioning with particular emphasis on nutrients. *Interciencia* 3:223–232.

Heyligers, P. C. 1963. Vegetation and soil of a white-sand savanna in Suriname. In J. Lanjouw and S. Versteegh, eds., *The Vegetation of Suriname*, vol. III. Amsterdam, Van Eedenfonds.

Hopkins, B. 1968. Vegetation of the Olokemeji Forest Reserve, Nigeria. V. The vegetation of the savanna site with special reference to its seasonal changes. *J. Ecol.* 56:97–115.

—— 1970. Vegetation of the Olokemeji Forest Reserve, Nigeria. VII. The plants of the savanna site with special reference to their seasonal growth. *J. Ecol.* 58:795–825.

Huber, O. 1974. Le savane neotropicali (Literature list) Rome, Instituto Italo-Latino Americano.

Hunt, W. F. 1970. The influence of leaf death on the rate of accumulation of green herbage during pasture regrowth. *J. App. Ecol.* 7:41–50.

Huntley, B. J. and B. H. Walker, eds. 1982. *Ecology of Tropical Savannas.* New York, Springer-Verlag.

Huttel, C. 1975. Recherches sur l'ecosystème de la forêt subéquatoriale de basse Côte d'Ivoire. IV. Estimation du bilan hydrique. *Terre Vie* 29:192–202.

Huttel, C., and F. Bernhard-Reversat. 1975. Recherches sur l'ecosystème de la forêt subéquatoriale de basse Côte d'Ivoire. V. Biomasse végétale et productivité primaire. Cycle de la matiére organique. *Terre Vie* 29:203–228.

Iwaki, H., I. Midorikawa, and K. Hogetsu. 1964. Studies on the productivity in Kirigamine grasslands, Central Japan. II. Seasonal changes in standing crop. *Bot. Mag. Tokyo* 77:447–457.

Jaeger, F. 1945. Zur Gliederung und Benenming des tropischen Graslandgürtels. *Verh. Naturf. Ges. Basel* 56:509–520.

Kamprath, E. J. 1972. Soil acidity and liming. In *Soils of the Humid Tropics*, pp. 136–149. Washington, D.C., National Academy of Sciences.

Kellman, M. 1979. Soil enrichment by neotropical savanna trees. *J. Ecol.* 67:565–577.

Kelly, J. M. 1975. Dynamics of root biomass in two eastern Tennessee old-field communities. *Am. Midl. Nat.* 94:54–61.

## 224    References

Kershaw, K. A. 1973. *Quantitative and Dynamic Ecology,* 2nd ed. London, Arnold.

Klinge, H. 1967. Podzol soils: a source of blackwater rivers in Amazonia. *Atas Simp. Biota Amazonica* 3:117–125.

—— 1977. Preliminary data on nutrient release from decomposing leaf litter in a neotropical rain forest. *Amazoniana* 6:193–202.

Klinge, H., E. Medina, and R. Herrera. 1977. Studies on the ecology of Amazon Caatinga forest in southern Venezuela. *Acta Cient. Venez.* 28:270–276.

Kucera, C. L., and J. H. Ehrenreich. 1962. Some effects of annual burning on central Missouri prairie. *Ecology* 43:334–336.

Kucera, C. L., R. C. Dahlman and R. Koelling. 1967. Total net productivity and turnover on an energy basis for tallgrass prairie. *Ecology* 48:536–541.

Labouriau, L. G. 1966. Revisao da situação da ecologia vegetal nos Cerrados. In L. G. Labouriau, ed., *II Simpósio sobre o Cerrado,* 5–38. An. Acad. Brasil. Cienc. 38 (suppl.).

Lamotte, M. 1975. The structure and function of a tropical savanna ecosystem. In E. Medina, and F. Golley, eds., *Tropical Ecological Systems,* 179–222. New York, Springer Verlag.

Lamotte, M., and F. Bourlière. 1983. Energy flow and nutrient cycling in tropical savannas. In F. Bourlière, ed., *Tropical Savannas,* pp. 583–603. Amsterdam, Elsevier.

Lanjouw, J. 1936. Studies on the vegetation of the Surinam savannas and swamps. *Ned. Kruindk. Arch.* 46:823–851.

Lauer, W. 1952. Humide und aride Jahreszeiten in Afrika und Sudamerika und ihre Beziehungen zu den Vegetationsgürteln. *Bonner Geogr. Abh.* 9.

Lecordier, C. 1974. Le climat. *Bull. liaison cherch. Lamto,* Número special 1974:45–103.

Lomnicki, A., E. Bandola, and K. Jankowska. 1968. Modification of the Wiegert-Evans method for estimation of net primary production. *Ecology* 49:147–149.

Loveless, A. R. 1961. A nutritional interpretation of sclerophylly based on differences in the chemical composition of sclerophyllous and mesophytic leaves. *Ann. Bot.* 25:168–184.

—— 1962. Further evidence to support a nutritional interpretation of sclerophylly. *Ann. Bot.* 26:551–561.

Lunt, O. R. 1972. Problems in nutrient availability and toxicity. In V. B. Youngner and C. McKell, eds., *The Biology and Utilization of Grasses,* pp. 271–277. New York, Academic Press.

Malavolta, E., J. R. Sarruge, and V. C. Bittencourt. 1977. Toxidez de Aluminio e de Manganes. In *IV Simpósio sobre o Cerrado,* pp. 275–301. São Paulo, Editóra da Universidade de São Paulo.

Malone, C. R. 1968. Determination of peak standing crop biomass of herbaceous shoots by the harvest method. *Am. Midl. Nat.* 79:429–435.

Markham, R. H., and A. J. Babbedge. 1979. Soil and vegetation catenas on the forest-savanna boundary in Ghana. *Biotropica* 11:224–234.

McColl, J. G. 1970. Properties of some natural waters in a tropical wet forest of Costa Rica. *BioScience* 20:1096–1100.

McGinnis, J. T., F. B. Golley, R. G. Clements, G. I. Child, and M. J. Diever. 1969. Elemental and hydrological budgets of the Panamanian moist forest. *BioScience* 19:697–700.

Medina, E. 1967. Intercambio gaseoso de árboles de las sabanas de *Trachypogon* en Venezuela. *Bol. Soc. Ven. Cienc. Nat.* 111:56–69.

——— 1978. Productivity and ecophysiological aspects of grassland ecosystems. (Unpublished paper, 30 pp.)

Medina, E., J. Silva, and E. Castellano. 1969. Variaciones estacionales del crecimiento y la respiración foliar de algunas plantas leñosas de las sabanas de *Trachypogon*. *Bol. Soc. Venez. Cienc. Nat.* 115–116:67–82.

Medina, E., A. Mendoza, and R. Montes. 1977. Balance nutricional y producción de materia orgánica en las sabanas de *Trachypogon* de Calabozo, Venezuela. *Bol. Soc. Venez. Cienc. Nat.* 134:101–120.

Medina, E., and G. Sarmiento. 1979. Tropical grazing land ecosystems of Venezuela. I. Ecophysiological studies in the *Trachypogon* savanna (central llanos). In *Tropical Grazing Land Ecosystems,* pp. 612–629. Paris, UNESCO.

Meiklejohn, J. 1962. Microbiology of the nitrogen cycle in some Ghana soils. *Emp. J. Exp. Agric.* 30:115–126.

Menaut, J. C. 1971. Etude de quelques peuplements ligneux d'une savane guineenne de Côte d'Ivoire. Ph.D. diss., University of Paris.

Menaut, J. C., and J. Cesar. 1979. Structure and primary productivity of Lamto savannas, Ivory Coast. *Ecology* 60:1197–1210.

Milner, C., and R. Elfyn Hughes. 1968. *Methods for the Measurement of the Primary Production of Grassland.* IBP Handbook No. 6. Oxford, Blackwell.

Monasterio, M. 1968. Observations sur les rythmes annuels de la savane tropicale des "Llanos" du Venezuela. Ph.D. diss., University of Montpellier.

——— 1970. Ecología de las sabanas de America tropical. II. Caracterización ecológica del clima en los llanos de Calabozo, Venezuela. *Revista Geográfica* 21:5–38.

Monasterio, M., and G. Sarmiento. 1968. Análisis ecológico y fitosociológico de la sabana en la Estación Biólogica de los llanos. *Bol. Soc. Venez. Cienc. Nat.* 113–114:477–524.

——— 1976. Phenological strategies in species of seasonal savanna and semi-deciduous forest in the Venezuelan Llanos. *J. Biogeogr.* 3:325–355.

Monsi, M. 1968. Mathematical models of plant communities. In F. E. Eckardt, ed., *Functioning of Terrestrial Ecosystems at the Primary Production Level,* pp. 131–148. Paris, UNESCO.

Monsi, M., and T. Saeki. 1953. Uber den Lichtfaktor in den Pflanzen-Gesellschaften und seine Bedeutung fur die Stoffproduktion. *Jap. J. Bot.* 14:22–52.

Montes, R., and E. Medina. 1977. Leaf nutrient content and ecological behaviour of trees and grasses in the *Trachypogon* savannas of Venezuela. *Geo-Eco-Trop.* 1:295–307.

Moore, A. W. 1963. Nitrogen fixation in latosolic soil under grass. *Plant & Soil* 19:127–138.

Moureaux, C., and G. Boquel. 1973. Microbiologie des sols ferralitiques. In Boissezon et al., *Les Sols Ferralitiques. Tome IV. La Matiére Organique et la Vie dans les Sols Ferraltiques*, pp. 67–106. Paris, ORSTOM.

Murphy, P. G. 1975. Net primary productivity in tropical terrestral ecosystems. In H. Lieth and R. H. Whittaker, eds., *Primary Productivity of the Biosphere*, 217–231. Berlin, Springer Verlag.

Nye, P. H. 1961. Organic matter and nutrient cycles under moist tropical forest. *Pl. & Soil* 13:333–346.

Nye, P. H., and D. J. Greenland. 1960. *The Soil under Shifting Cultivation.* London, Tech. Commonw. Bur. Soils 51.

Odum, E. P. 1971. *Fundamentals of Ecology.* 3rd ed. Philadelphia, Saunders.

Philip, J. R. 1966. Plant water relations: some physical aspects. *Ann. Rev. Plant Physiol.* 17:245–268.

Poissonet, J., and J. Cesar. 1972. Structure spécifique de la strate herbacée dans la savane à palmier ronier de Lamto (Côte d'Ivoire). *Ann. Univ. Abidjan, Serie E* (Ecologie) 4:577–601.

Pratt, D. J., P. J. Greenway, and M. D. Gwynne. 1966. A classification of east African rangelands, with an appendix on terminology. *J. Appl. Ecol.* 3:369–382.

Pulle, A. A. 1906. *An Enumeration of the Vascular Plants known from Surinam, together with Their Distribution and Synonymy.* Leiden, E. J. Brill.

Rachid, M. 1947. Transpiração e sistemas subterraneos da vegetacão de verao dos campos cerrados de Emas. *Bol. Fac. Fil. Cienc. Letr. U.S.P. 80, Botánica* 5:1–35.

Ranzani, G. 1963. Solos do cerrado. In *Simpósio sobre o Cerrado*, pp. 53–92. São Paulo, Editóra da Universidade de São Paulo.

——— 1971. Solos do cerrado no Brasil. In *III Simpósio sobre o Cerrado*, pp. 26–43. São Paulo, Editóra da Universidade de São Paulo.

Ratter, J. A., P. W. Richards, G. Argent, and D. R. Gifford. 1973. Observations on the vegetation of northeastern Mato Grosso. I. The woody vegetation types of the Xavantina-Cachimbo expedition area. *Phil. Trans. R. Soc. B Biol. Sci.* 266:449–492.

Rawitscher, F., M. G. Ferri, and M. Rachid. 1943. Profundidade dos solos e vegetação em campos cerrados do Brasil meridional. *An. Acad. Brasil, Cienc.* 15:267–294.

Rawitscher, F. and M. Rachid. 1946. Troncos subterráneos de plantas brasileiras. *An. Acad. Brasil, Cienc.* 18:261–280.

Rawitscher, F. 1948. The water economy of the vegetation of the "campos cerrados" in southern Brazil. *J. Ecol.* 36:238–268.

Rham, P. de. 1969. L'azote dans quelques forêts, savanes et terrains de culture d'Afrique tropicale humide (Côte d'Ivoire). Ph.D. diss., University of Lausanne.

——— 1973. Recherches sur la minéralisation de l'azote dans les sols des savanes de Lamto (Côte d'Ivoire). *Rev. Ecol. Biol. Sol.* 10:169–196.

San José, J. J., and M. Farinas. 1971. Estudio sobre los cambios de la vegetación protegida de la quema y el pastoreo en la estación biológica de Los Llanos. *Bol. Soc. Venez. Cienc. Nat.* 119–120:136–144.

San José, J. J., and E. Medina. 1975. Effect of fire on organic matter production and water balance in a tropical savanna. In E. Medina and F. Golley, eds., *Tropical Ecological Systems*, pp. 251–264. New York, Springer Verlag.

——— 1976. Organic matter production in the *Trachypogon* savanna at Calabozo, Venezuela. *Trop. Ecol.* 17:113–124.

——— 1977. Producción de materia orgánica en la sabana de *Trachypogon*, Calabozo, Venezuela. *Bol. Soc. Venez. Cienc. Nat.* 134:75–100.

San José, J. J., and J. García Miragaya. 1981. Factores ecológicos operacionales en la producción de materia orgánica de las sabanas de *Trachypogon*. *Bol. Soc. Venez. Cienc. Nat.* 139:347–374.

San José, J. J., F. Berrade, and J. Ramirez. 1982. Seasonal changes of growth, mortality and disappearance of belowground root biomass in the *Trachypogon* savanna grass. *Acta Oecologica, Oecol. Plant.* 3:347–358.

Sarmiento, G. 1983. The savannas of tropical America. In F. Bourlière, ed., *Tropical Savannas*, pp. 245–288. Amsterdam, Elsevier.

Sarmiento, G. (in press). Patterns of specific and phenological diversity in the grass community of the Venezuelan tropical savannas. *J. Biogeogr.*

Sarmiento, G., and M. Monasterio. 1969. Corte ecológico del Estado Guárico. *Bol. Soc. Venez. Cienc. Nat.* 115–116:83–106.

——— 1971. Ecología de las sabanas de América tropical. I. Análisis macro ecológico de los Llanos de Calabozo, Venezuela. *Cuadernos Geográficos* 4:1–126.

——— 1975. A critical consideration of the environmental conditions associated with the occurrence of savanna ecosystems in tropical America. In E. Medina and F. B. Golley, eds., *Tropical Ecological Systems*, pp. 223–250. New York, Springer Verlag.

——— 1983. Life forms and phenology. In F. Bourlière, ed., *Tropical Savannas*, pp. 79–108. Amsterdam, Elsevier.

Sarmiento, G., M. Monasterio, and J. Silva. 1971. Reconocimiento ecológico de los Llanos Occidentales. I. Las unidades ecológicas regionales. *Acta Cient. Venez.* 22:52–60.

Sarmiento, G., and M. Vera. 1977. La marcha anual del agua en el suelo en sabanas y bosques tropicales de los Llanos de Venezuela. *Agron. Trop.* 27:629–649.

——— 1979. Composición, estructura, biomasa y producción de diferentes sabanas en los Llanos de Venezuela. *Bol. Soc. Venez. Cienc. Nat.* 136:5–41.

Schimper, A. F. W., and F. C. von Faber. 1935. *Pflanzengeographie auf physiologischer Grundlage.* 3rd ed., Jena, G. Fisher.

Shankar, V., K. A. Shankarnarayan, and P. Rai. 1973. Primary production, energetics and nutrient cycling in *Sehima-Heteropogon* grassland. I. Seasonal variations in composition, standing crop and net production. *Trop. Ecol.* 14:238–251.

Silva, J., and G. Sarmiento. 1976. La composición de las sabanas en Barinas en relación con las unidades edáficas. *Acta Cient. Venez.* 27:68–78.

—— 1976. Influencia de factores edáficos en la diferenciación de las sabanas. Análisis de componentes principales y su interpretación ecológica. *Acta Cient. Venez.* 27:141–147.

Silva, J. F., and M. Ataroff, 1982. Fenología, estrategias reproductivas y coexistencia en especies de gramíneas de una sabana tropical. *Acta Cient. Venez.* 33 (supl. 1):99.

Sims, P. L. and J. S. Singh. 1978a. The structure and function of ten western North American grasslands. II. Intra-seasonal dynamics in primary producer compartments. *J. Ecol.* 66:547–572.

—— 1978b. The structure and function of ten western North American grasslands. III. Net primary production, turnover, and efficiencies of energy capture and water use. *J. Ecol.* 66:573–597.

—— 1978c. The structure and function of ten western North American grasslands. IV. Compartmental transfers and energy flow within the ecosystem. *J. Ecol.* 66:983–1009.

Singh, J. S., and P. S. Yadava. 1974. Seasonal variation in composition, plant biomass, and net primary productivity of a tropical grassland at Kurukshetra, India. *Ecological Monographs* 44:351–376.

Singh, J. S., W. K. Lavenroth, and R. K. Steinhorst. 1975. Review and assessment of various techniques for estimating net aerial production in grasslands from harvest data. *Bot. Rev.* 41:181–232.

Sinha, N. P, K. 1968. Geomorphic evolution of the Northern Rupununi Basin, Guyana. Montreal, McGill University Savanna Research Series No. 11.

Sioli, H. 1967. Studies in Amazonian waters. *Atas Simp. Biota Amazonica* 3:9–50.

—— 1975. Tropical rivers as expressions of their terrestrial environments. In E. Medina and F. B. Golley, eds., *Tropical Ecological Systems*, pp. 275–288. New York, Springer Verlag.

Soares, W. V., E. Lobato, E. González, and G. C. Naderman. 1974. Encalado de los suelos del Cerrado Brasileño. In E. Bornemisza and A. Alvarado, eds., *Manejo de Suelos en la América Tropical*, pp. 287–303. Raleigh, N.C., Soil Sc. Dept., North Carolina State University Press.

Stanyukovich, K. V. 1970. An attempt to classify world plant communities on the basis of their ecological rhythm. *Ekologiya* 1:18–26.

Stark, N. 1970. The nutrient content of plants and soils from Brazil and Surinam. *Biotropica* 2:51–60.

—— 1971a. Nutrient cycling I. Nutrient distribution in some Amazonian soils. *Trop. Ecol.* 12:24–50.

—————— 1971b. Nutrient cycling. II. Nutrient distribution in an Amazonian vegetation. *Trop. Ecol.* 12:177–201.

Steyermark, J., and C. Brewer-Carias. 1976. La vegetación de la cima del macizo de Jaua. *Bol. Soc. Venez. Cienc. Nat.* 132–33:179–405.

Torres, A., and F. Meinzer. 1982. Estudio de algunos aspectos de la ecofisiología de tres gramíneas en la sabana estacional. *Acta Cient. Venez.* 33 (supl. 1):114.

Tricart, J., and A. Millies-Lacroix. 1962. Les terraces quaternaires des Andes vénézuéliennes. *Bull. Soc. Geol. France.* 7éme serie, IV:201–218.

Troll, C. 1950. Savannentypen und das Problem der Primarsavannen. *Proc. 7th Int. Bot. Congress, Stockholm,* pp. 670–674. Stockholm, Int. Bot. Congress.

Trómpiz, I., and J. F. Silva. 1982. Estudio comparativo de algunos microhabitats en una sabana estacional. *Acta Cient. Venez.* 33 (suppl. 1):93.

Troughton, A. 1957. The underground organs of herbage grasses. *Comm. Agric. Bur., Bull.* 44:1–163.

UNESCO. 1979. *Tropical grazing land ecosystems, a state-of-knowledge report prepared by Unesco/UNEP/FAO.* Natural Resources Research XVI. Paris, UNESCO.

Van Donselaar-Ten Bokkel Huinink, W. A. E. 1966. Structure, root systems, and periodicity of savanna plants and vegetation in northern Surinam. *Wentia* 17:1–162.

Van Donselaar, J. 1965. An ecological and phytogeographic study of northern Surinam savannas. *Wentia* 14:1–163.

Varshney, C. K. 1972. Productivity of Delhi grasslands. In P. M. Golley and F. B. Golley, eds., *Tropical Ecology,* pp. 27–42. Athens, Ga., University of Georgia Press.

Vera, M. 1977. Producción de hojarasca y retorno de nutrientes al suelo en una sabana arbolada. *IV Simp. Intern. Ecol. Trop., Panamá, Resúmenes:* 128–130.

Waibel, L. 1948. Vegetation and land use in the Planalto Central of Brazil. *Geog. Rev.* 38:529–554.

Walter, H. 1971. *Ecology of Tropical and Subtropical Vegetation.* Edinburgh, Oliver and Boyd.

Warming, E. 1892, rpt. 1973. *Lagoa Santa.* São Paulo, Editóra da Universidade de São Paulo.

Weaver, J. E., and F. E. Clements. 1929, rpt. 1946. *Plant Ecology.* New York, McGraw-Hill; Buenos Aires, Acme Agency.

Weaver, J. E., and E. Zink. 1946. Annual increase of underground materials in three range grasses. *Ecology* 27:115–127.

Went, F. W., and N. Stark. 1968. The biological and mechanical role of soil fungi. *Proc. Nat. Acad. Sci.* 60:497–504.

Whittaker, R. H., and P. L. Marks. 1975. Methods of assessing terrestrial productivity. In H. Lieth and R. H. Whittaker, eds., *Primary Productivity of the Biosphere,* pp. 55–118. Berlin, Springer Verlag.

Wiegert, R. G., and F. C. Evans. 1964. Primary production and the disappearance of dead vegetation in an old field in southeastern Michigan. *Ecology* 45:49–63.

Williams, W. A., R. S. Loomis, and P. de T. Alvim. 1972. Environments of evergreen rain forests on the lower Rio Negro, Brazil. *Trop. Ecol.* 13:65–78.

Zinck, A. 1970. *Aplicación de la geomorfología al levantamiento de suelos en zonas aluviales.* Barcelona, Venezuela, División de Edafología, M.O.P.

Zinck, A., and P. Stagno, 1966. *Estudio Edafológico de la Zona Rio Santo Domingo-Rio Paguey, Estado Barinas.* Guanare, División de Edafología, M.O.P.

Zinck, A., and P. L. Urriola. 1968. *Estudio Edafológico del Valle del Rio Guarapiche, Estado Monagas.* Caracas, División de Edafología, M.O.P.

# Index